Project-Based Inquiry Science™

DIGGING IN

Janet L. Kolodner

Joseph S. Krajcik

Daniel C. Edelson

Brian J. Reiser

Mary L. Starr

IT's ABOUT TIME®

333 North Bedford Road, Mount Kisco, NY 10549
Phone (914) 273-2233 Fax (914) 206-6444
www.IAT.com

Program Components

Student Edition
Teacher's Planning Guide
Durable Equipment Kit
Consumable Equipment Kit

All student activities in this textbook have been designed to be as safe as possible, and have been reviewed by professionals specifically for that purpose. As well, appropriate warnings concerning potential safety hazards are included where applicable to particular activities. However, responsibility for safety remains with the student, the classroom teacher, the school principal, and the school board.

Printed and bound in the United States of America.

The 3rd printing of *Digging In* was completed
by Courier Kendallville, Inc.in Kendallville, IN in June 2013.

ISBN 978-1-58591-617-7

1st Printing

1 2 3 4 5 6 16 15 14 13

This project was supported, in part, by the ***National Science Foundation*** under grant nos. 0137807, 0527341, 0639978.
Opinions expressed are those of the authors and not necessarily those of the National Science Foundation.

PBIS Principal Investigators

Janet L. Kolodner is a Regents' Professor in the School of Interactive Computing in the Georgia Institute of Technology's College of Computing. Since 1978, her research has focused on learning from experience, both in computers and in people. She pioneered the Artificial Intelligence method called *case-based reasoning*, providing a way for computers to solve new problems based on their past experiences. Her book, *Case-Based Reasoning*, synthesizes work across the case-based reasoning research community from its inception to 1993.

Since 1994, Dr. Kolodner has focused on the applications and implications of case-based reasoning for education. In her approach to science education, called Learning by Design™ (LBD), students learn science while pursuing design challenges. Dr. Kolodner has investigated how to create a culture of collaboration and rigorous science talk in classrooms, how to use a project challenge to promote focus on science content, and how students learn and develop when classrooms function as learning communities. Currently, Dr. Kolodner is investigating how to help young people come to think of themselves as scientific reasoners. Dr. Kolodner's research results have been widely published, including in *Cognitive Science, Design Studies,* and the *Journal of the Learning Sciences.*

Dr. Kolodner was founding Director of Georgia Tech's EduTech Institute, served as coordinator of Georgia Tech's Cognitive Science program for many years, and is founding Editor in Chief of the *Journal of the Learning Sciences*. She is a founder of the International Society for the Learning Sciences, and she served as its first Executive Officer. She is a fellow of the American Association of Artificial Intelligence.

Joseph S. Krajcik is a Professor of Science Education and Associate Dean for Research in the School of Education at the University of Michigan. He works with teachers in science classrooms to bring about sustained change by creating classroom environments in which students find solutions to important intellectual questions that subsume essential curriculum standards and use learning technologies as productivity tools. He seeks to discover what students learn in such environments, as well as to explore and find solutions to challenges that teachers face in enacting such complex instruction. Dr. Krajcik has authored and co-authored over 100 manuscripts and makes frequent presentations at international, national, and regional conferences that focus on his research, as well as presentations that translate research findings into classroom practice. He is a fellow of the American Association for the Advancement of Science and served as president of the National Association for Research in Science Teaching. Dr. Krajcik co-directs the Center for Highly Interactive Classrooms, Curriculum and Computing in Education at the University of Michigan and is a co-principal investigator in the Center for Curriculum Materials in Science and The National Center for Learning and Teaching Nanoscale Science and Engineering. In 2002, Dr. Krajcik was honored to receive a Guest Professorship from Beijing Normal University in Beijing, China. In winter 2005, he was the Weston Visiting Professor of Science Education at the Weizmann Institute of Science in Rehovot, Israel.

Daniel C. Edelson is Vice President for Education and Children's Programs at the National Geographic Society. Previously, he was the director of the Geographic Data in Education (GEODE) Initiative at Northwestern University, where he led the development of Planetary Forecaster and Earth Systems and Processes. Since 1992, Dr. Edelson has directed a series of projects exploring the use of technology as a catalyst for reform in science education and has led the development of a number of software environments for education. These include My World GIS, a geographic information system for inquiry-based learning, and WorldWatcher, a data visualization and analysis system for gridded geographic data. Dr. Edelson is the author of the high school environmental science text, *Investigations in Environmental Science: A Case-Based Approach to the Study of Environmental Systems*. His research has been widely published, including in the *Journal of the Learning Sciences*, the *Journal of Research on Science Teaching*, *Science Educator,* and *The Science Teacher*.

Brian J. Reiser is a Professor of Learning Sciences in the School of Education and Social Policy at Northwestern University. Professor Reiser served as chair of Northwestern's Learning Sciences Ph.D. program from 1993, shortly after its inception, until 2001. His research focuses on the design and enactment of learning environments that support students' inquiry in science, including both science curriculum materials and scaffolded software tools. His research investigates the design of learning environments that scaffold scientific practices, including investigation, argumentation, and explanation; design principles for technology-infused curricula that engage students in inquiry projects; and the teaching practices that support student inquiry. Professor Reiser also directed BGuILE (Biology Guided Inquiry Learning Environments) to develop software tools for supporting middle school and high school students in analyzing data and constructing explanations with biological data. Reiser is a co-principal investigator in the NSF Center for Curriculum Materials in Science. He served as a member of the NRC panel authoring the report Taking Science to School.

Mary L. Starr is a Research Specialist in Science Education in the School of Education at the University of Michigan. She collaborates with teachers and students in elementary and middle school science classrooms around the United States who are implementing *Project-Based Inquiry Science*. Before joining the PBIS team, Dr. Starr created professional learning experiences in science, math, and technology, designed to assist teachers in successfully changing their classroom practices to promote student learning from coherent inquiry experiences. She has developed instructional materials in several STEM areas, including nanoscale science education, has presented at national and regional teacher education and educational research meetings, and has served in a leadership role in the Michigan Science Education Leadership Association. Dr. Starr has authored articles and book chapters, and has worked to improve elementary science teacher preparation through teaching science courses for pre-service teachers and acting as a consultant in elementary science teacher preparation. As part of the PBIS team, Dr. Starr has played a lead role in making units cohere as a curriculum, in developing the framework for PBIS Teacher's Planning Guides, and in developing teacher professional development experiences and materials.

Acknowledgements

Three research teams contributed to the development of *Project-Based Inquiry Science (PBIS)*: a team at the Georgia Institute of Technology headed by Janet L. Kolodner, a team at Northwestern University headed by Daniel Edelson and Brian Reiser, and a team at the University of Michigan headed by Joseph Krajcik and Ron Marx. Each of the PBIS units was originally developed by one of these teams and then later revised and edited to be a part of the full three-year middle-school curriculum that became PBIS.

PBIS has its roots in two educational approaches, Project-Based Science and Learning by Design™. Project-Based Science suggests that students should learn science through engaging in the same kinds of inquiry practices scientists use, in the context of scientific problems relevant to their lives and using tools authentic to science. Project-Based Science was originally conceived in the hi-ce Center at the University of Michigan, with funding from the National Science Foundation. Learning by Design™ derives from Problem-Based Learning and suggests sequencing, social practices, and reflective activities for promoting learning. It engages students in design practices, including the use of iteration and deliberate reflection. LBD was conceived at the Georgia Institute of Technology, with funding from the National Science Foundation, DARPA, and the McDonnell Foundation.

The development of the integrated PBIS curriculum was supported by the National Science Foundation under grants nos. 0137807, 0527341, and 0639978. Any opinions, findings and conclusions, or recommendations expressed in this material are those of the authors and do not necessarily reflect the views of the National Science Foundation.

PBIS Team

Principal Investigator
Janet L. Kolodner

Co-Principal Investigators
Daniel C. Edelson
Joseph S. Krajcik
Brian J. Reiser

NSF Program Officer
Gerhard Salinger

Curriculum Developers
Michael T. Ryan
Mary L. Starr

Teacher's Edition Developers
Rebecca M. Schneider
Mary L. Starr

Literacy Specialist
LeeAnn M. Sutherland

NSF Program Reviewer
Arthur Eisenkraft

Project Coordinator
Juliana Lancaster

External Evaluators
The Learning Partnership
Steven M. McGee
Jennifer Witers

The Georgia Institute of Technology Team

Project Director:
Janet L. Kolodner

Development of PBIS units at the Georgia Institute of Technology was conducted in conjunction with the Learning by Design™ Research group (LBD), Janet L. Kolodner, PI.

Lead Developers, Physical Science:
David Crismond
Michael T. Ryan

Lead Developer, Earth Science:
Paul J. Camp

Lead Ethnographer:
Barbara Fasse

Assessment and Evaluation:
Jackie Gray
Daniel Hickey
Jennifer Holbrook
Laura Vandewiele

Project Pioneers:
JoAnne Collins
David Crismond
Joanna Fox
Alice Gertzman
Mark Guzdial
Cindy Hmelo-Silver
Douglas Holton
Roland Hubscher
N. Hari Narayanan
Wendy Newstetter
Valery Petrushin
Kathy Politis
Sadhana Puntambekar
David Rector
Janice Young

The Northwestern University Team

Project Directors:
Daniel Edelson
Brian Reiser

Lead Developer, Biology:
David Kanter

Lead Developers, Earth Science:
Jennifer Mundt Leimberer
Darlene Slusher

Development of PBIS units at Northwestern was conducted in conjunction with:

The Center for Learning Technologies in Urban Schools (LeTUS) at Northwestern, and the Chicago Public Schools
Clifton Burgess, PI
for Chicago Public Schools;
Louis Gomez, PI.

The BioQ Collaborative
David Kanter, PI.

The Biology Guided Inquiry Learning Environments (BGuILE) Project
Brian Reiser, PI.

The Geographic Data in Education (GEODE) Initiative
Daniel Edelson, Director

The Center for Curriculum Materials in Science at Northwestern
Daniel Edelson,
Brian Reiser,
Bruce Sherin, PIs.

The University of Michigan Team

Project Directors:
Joseph Krajcik
Ron Marx

Literacy Specialist:
LeeAnn M. Sutherland

Project Coordinator:
Mary L. Starr

Development of PBIS units at the University of Michigan was conducted in conjunction with:

The Center for Learning Technologies in Urban Schools (LeTUS)
Phyllis Blumenfeld,
Barry Fishman,
Joseph Krajcik,
Ron Marx,
Elliot Soloway, PIs.

The Detroit Public Schools
Juanita Clay-Chambers
Deborah Peek-Brown

The Center for Highly Interactive Computing in Education (hi-ce)
Phyllis Blumenfeld,
Barry Fishman,
Joseph Krajcik,
Ron Marx,
Elizabeth Moje,
Elliot Soloway,
LeeAnn Sutherland, PIs.

Field-Test Teachers

National Field Test

Tamica Andrew
Leslie Baker
Jeanne Bayer
Gretchen Bryant
Boris Consuegra
Daun D'Aversa
Candi DiMauro
Kristie L. Divinski
Donna M. Dowd
Jason Fiorito
Lara Fish
Christine Gleason
Christine Hallerman
Terri L. Hart-Parker
Jennifer Hunn
Rhonda K. Hunter
Jessica Jones
Dawn Kuppersmith
Anthony F. Lawrence
Ann Novak
Rise Orsini
Tracy E. Parham
Cheryl Sgro-Ellis
Debra Tenenbaum
Sarah B. Topper
Becky Watts
Debra A. Williams
Ingrid M. Woolfolk
Ping-Jade Yang

New York City Field Test

Several sequences of PBIS units have been field- tested in New York City under the leadership of Whitney Lukens, Staff Developer for Region 9, and Greg Borman, Science Instructional Specialist, New York City Department of Education

6th Grade

Norman Agard
Tazinmudin Ali
Heather Guthartz Aniba
Asher Arzonane
Asli Aydin
Shareese Blakely
John J. Blaylock
Joshua Blum
Tsedey Bogale
Filomena Borrero
Zachary Brachio
Thelma Brown
Alicia Browne-Jones
Scott Bullis
Maximo Cabral
Lionel Callender
Matthew Carpenter
Ana Maria Castro
Diane Castro
Anne Chan
Ligia Chiorean
Boris Consuegra
Careen Halton Cooper
Cinnamon Czarnecki
Kristin Decker
Nancy Dejean
Gina DiCicco
Donna Dowd
Lizanne Espina
Joan Ferrato
Matt Finnerty
Jacqueline Flicker
Helen Fludd
Leigh Summers Frey
Helene Friedman-Hager
Diana Gering
Matthew Giles
Lucy Gill
Steven Gladden
Greg Grambo
Carrie Grodin-Vehling
Stephan Joanides
Kathryn Kadei
Paraskevi Karangunis
Cynthia Kerns
Martine Lalanne
Erin Lalor
Jennifer Lerman
Sara Lugert
Whitney Lukens
Dana Martorella
Christine Mazurek
Janine McGeown
Chevelle McKeever
Kevin Meyer
Jennifer Miller
Nicholas Miller
Diana Neligan
Caitlin Van Ness
Marlyn Orque
Eloisa Gelo Ortiz
Gina Papadopoulos
Tim Perez
Albertha Petrochilos
Christopher Poli
Kristina Rodriguez
Nadiesta Sanchez
Annette Schavez
Hilary Sedgwitch
Elissa Seto
Laura Shectman
Audrey Shmuel
Katherine Silva
Ragini Singhal
C. Nicole Smith
Gitangali Sohit
Justin Stein
Thomas Tapia
Eilish Walsh-Lennon
Lisa Wong
Brian Yanek
Cesar Yarleque
David Zaretsky
Colleen Zarinsky

7th Grade

Mayra Amaro
Emmanuel Anastasiou
Cheryl Barnhill
Bryce Cahn
Ligia Chiorean
Ben Colella
Boris Consuegra
Careen Halton Cooper
Elizabeth Derse
Urmilla Dhanraj
Gina DiCicco
Lydia Doubleday
Lizanne Espina
Matt Finnerty
Steven Gladden
Stephanie Goldberg
Nicholas Graham
Robert Hunter
Charlene Joseph
Ketlynne Joseph
Kimberly Kavazanjian
Christine Kennedy
Bakwah Kotung
Lisa Kraker
Anthony Lett
Herb Lippe
Jennifer Lopez
Jill Mastromarino
Kerry McKie
Christie Morgado
Patrick O'Connor
Agnes Ochiagha
Tim Perez
Nadia Piltser
Chris Poli
Carmelo Ruiz
Kim Sanders
Leslie Schiavone
Ileana Solla
Jacqueline Taylor
Purvi Vora
Ester Wiltz
Carla Yuille
Marcy Sexauer Zacchea
Lidan Zhou

8th Grade

Emmanuel Anastasio
Jennifer Applebaum
Marsha Armstrong
Jenine Barunas
Vito Cipolla
Kathy Critharis
Patrecia Davis
Alison Earle
Lizanne Espina
Matt Finnerty
Ursula Fokine
Kirsis Genao
Steven Gladden
Stephanie Goldberg
Peter Gooding
Matthew Herschfeld
Mike Horowitz
Charlene Jenkins
Ruben Jimenez
Ketlynne Joseph
Kimberly Kavazanjian
Lisa Kraker
Dora Kravitz
Anthony Lett
Emilie Lubis
George McCarthy
David Mckinney
Michael McMahon
Paul Melhado
Jen Miller
Christie Morgado
Ms. Oporto
Maria Jenny Pineda
Anastasia Plaunova
Carmelo Ruiz
Riza Sanchez
Kim Sanders
Maureen Stefanides
Dave Thompson
Matthew Ulmann
Maria Verosa
Tony Yaskulski

Digging In

Digging In was originally developed at the Georgia Institute of Technology as part of the Learning by Design™ initiative and was titled *Earth Science: Digging In.*

Digging In

PBIS Editorial Team:

Michael T. Ryan
Victoria Deneroff

Georgia Institute of Technology Team

Project Director

Janet L. Kolodner

Lead Authors

Paul Camp
Jennifer Holbrook

Contributing Authors

David Crismond
Mike Ryan
Jennifer Turns

Formative Development

David Crismond
Joanna Fox
Jackie Gray
Cami Heck
Jennifer Holbrook
Susan McClendon
Kristine Nagel
Lindy Wine
Janice Young

Pilot Teachers

Barbara Blasch
Audrey Daniel
Emily Dickson
Carmen Dillard
Yvette Fernandez
Joyce Gamble
Dorothy Hicks
Daphne Islam-Gordon
Rudo Kashiri
Marni Klein
Toni Laman
Paige Lefont
Susan McClendon
Bernie Moore
Kelly Rowsey
Mike Ryan
Beth Smith
Delilah Springer
Lindy Wine
Avis Winfield
Mary Winn
Ann Yergin

The development of *Earth Science: Digging In* was supported by the National Science Foundation under grant nos. 9553583, 9818828 and 0208059 and by grants from the McDonell Foundation, the BellSouth Foundation, the Woodruff Foundation, and the Georgia Tech Foundation. Any opinions, findings, and conclusions or recommendations expressed in this material are those of the authors and do not necessarily reflect the views of the National Science Foundation.

Table of Contents

Welcome to Project-Based Inquiry Science!

Dear Students,

This year, you will be learning the way scientists learn. You will explore interesting questions and challenges. You will learn new things. You will also learn exciting, new ways to think about the world around you.

Scientists learn as they are trying to answer a big question or solve a big challenge. To help them work on these big questions or challenges, scientists break them into smaller ones. For each smaller question or challenge, they read what other scientists already know, and they investigate, explore, gather evidence, and form explanations. This way, scientists build new knowledge as they answer these smaller questions. Then they use what they have learned to try to answer the big question or solve the big challenge. Along the way, scientists share what they have learned with other scientists. These other scientists can then use this new knowledge to address other questions and challenges.

Like scientists, you will be trying to answer big questions and solve big challenges this year. You will break these into smaller questions or challenges. For each smaller question or challenge, you'll read, investigate, explore, gather evidence, and form explanations. As you do these things, you will share what you are learning and work closely with your classmates. You and your classmates will help each other learn and successfully answer each unit's question or solve its challenge. At the end of each unit, you'll answer the big question or address the big challenge based on what you've learned. And you will have learned a lot!

PBIS was designed to support you as you become a student scientist. In fact, PBIS units were written by scientists who study how people learn and who want to help you become the best scientist you can be. We used what we know about learning to design ways to help you answer big questions and solve big challenges. What you learn this year about science will help you learn science in the future. The way you learn to think about questions and challenges will help you learn other subjects, too.

Each year begins with a launcher unit. Launcher units help your class learn to work together, help you become familiar with the ways scientists think and have discussions, and introduce you to the activities and tools you'll use throughout PBIS.

Have fun being a student scientist!

Janet L. Kolodner

Janet Kolodner (for the whole PBIS team)

Introducing PBIS

What Do Scientists Do?

1) Scientists...address big challenges and big questions.

You will find many different kinds of *Big Challenges* and *Questions* in *PBIS* Units. Some ask you to think about why something is a certain way. Some ask you to think about what causes something to change. Some challenge you to design a solution to a problem. Most of them are about things that can and do happen in the real world.

Understand the Big Challenge or Question

As you get started with each Unit, you will do activities that help you understand the *Big Question* or *Challenge* for that Unit. You will think about what you already know that might help you, and you will identify some of the new things you will need to learn.

Project Board

The *Project Board* helps you keep track of your learning. For each challenge or question, you will use a *Project Board* to keep track of what you know, what you need to learn, and what you are learning. As you learn and gather evidence, you will record that on the *Project Board*. After you have answered each small question or challenge, you will return to the *Project Board* to record how what you've learned helps you answer the *Big Question* or *Challenge*.

Learning Sets

Each Unit is composed of a group of *Learning Sets*, one for each of the smaller questions that need to be answered to address the *Big Question* or *Challenge*. In each *Learning Set*, you will investigate and read to find answers to the *Learning Set's* question. You will also have a chance to share the results of your investigations with your classmates and work together to make sense of what you are learning. As you come to understand answers to the questions on the *Project Board*, you will record those answers and the evidence you've collected. At the end of each *Learning Set*, you will apply your knowledge to the *Big Question* or *Challenge*.

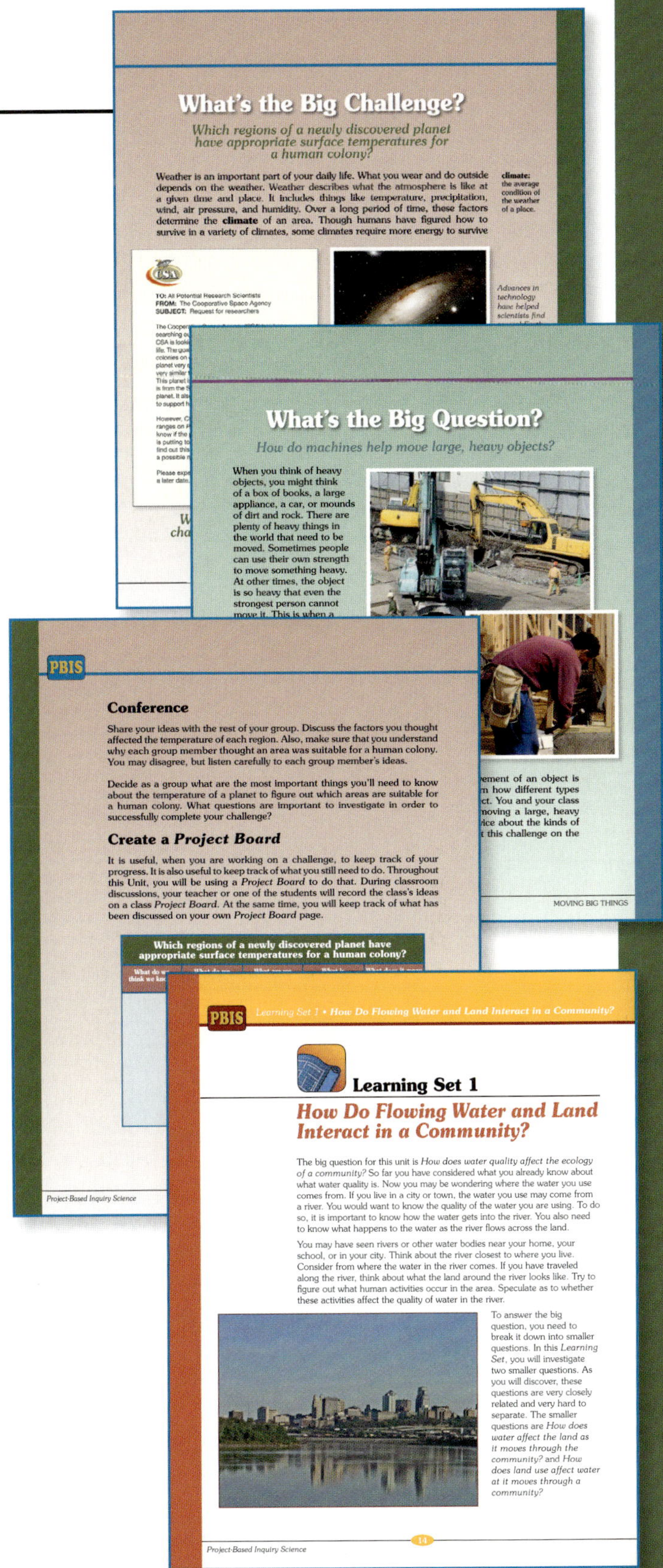

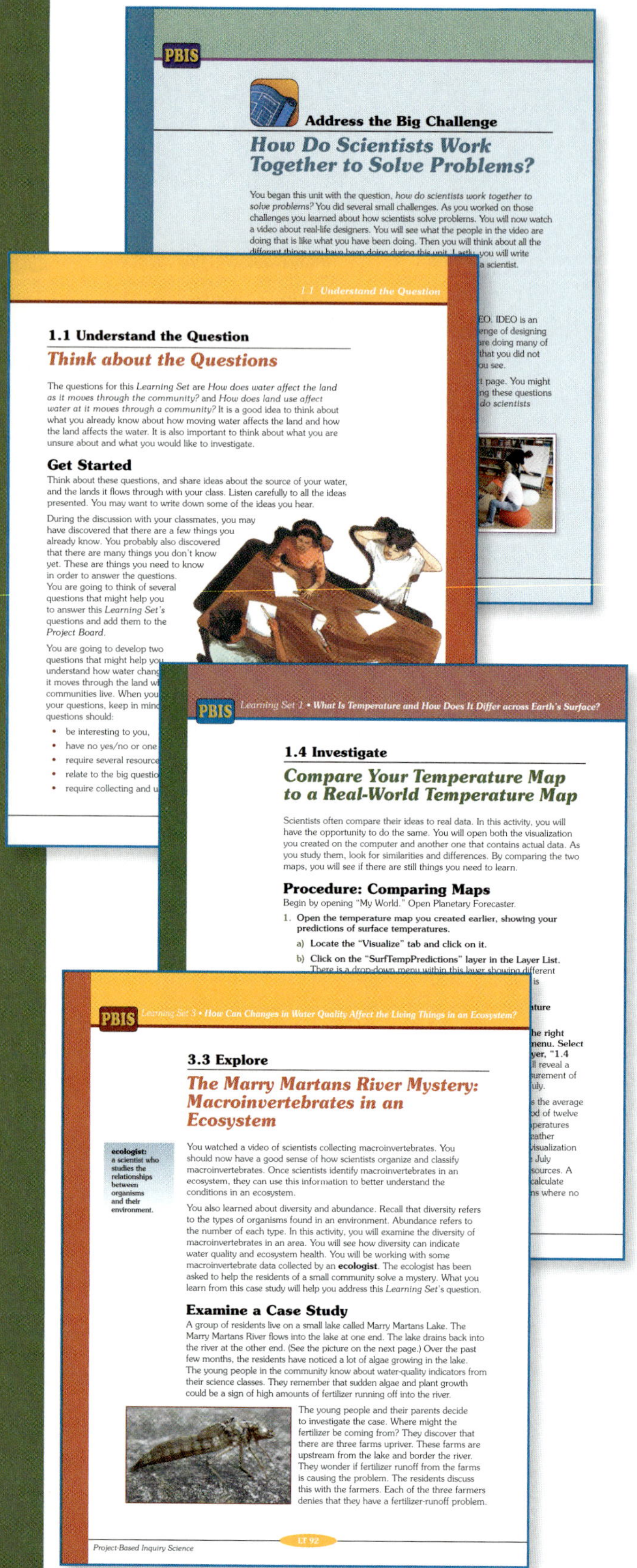

Answer the Big Question / Address the Big Challenge

At the end of each Unit, you will put everything you have learned together to tackle the *Big Question or Challenge*.

2) Scientists...address smaller questions and challenges.

What You Do in a Learning Set

Understanding the Question or Challenge

At the start of each *Learning Set*, you will usually do activities that will help you understand the *Learning Set's* question or challenge and recognize what you already know that can help you answer the question or achieve the challenge. Usually, you will visit the *Project Board* after these activities and record on it the even smaller questions that you need to investigate to answer a *Learning Set's* question.

Investigate/Explore

There are many different kinds of investigations you might do to find answers to questions. In the *Learning Sets*, you might

- design and run experiments;
- design and run simulations;
- design and build models;
- examine large sets of data.

Don't worry if you haven't done these things before. The text will provide you with lots of help in designing your investigations and in analyzing your data.

Read

Like scientists, you will also read about the science you are learning. You'll read a little bit before you investigate, but most of the reading you do will be to help you understand what you've experienced or seen in an investigation. Each time you read, the text will include *Stop and Think* questions after the reading. These questions will help you gauge how well you understand what you have read. Usually, the class will discuss the answers to *Stop and Think* questions before going on so that everybody has a chance to make sense of the reading.

Design and Build

When the *Big Challenge* for a Unit asks you to design something, the challenge in a *Learning Set* might also ask you to design something and make it work. Often, you will design a part of the thing you will design and build for the *Big Challenge.* When a *Learning Set* challenges you to design and build something, you will do several things:

- identify what questions you need to answer to be successful
- investigate to find answers to those questions
- use those answers to plan a good design solution
- build and test your design.

Because designs don't always work the way you want them to, you will usually do a design challenge more than once. Each time through, you will test your design. If your design doesn't work as well as you'd like, you will determine why it is not working and identify other things you need to investigate to make it work better. Then, you will learn those things and try again.

Explain and Recommend

A big part of what scientists do is explain, or try to make sense of why things happen the way they do. An explanation describes why something is the way it is or behaves the way it does. An explanation is a statement you make built from claims (what you think you know), evidence (from an investigation) that supports the claim, and science knowledge. As they learn, scientists get better at explaining. You'll see that you get better, too, as you work through the *Learning Sets.*

A recommendation is a special kind of claim—one where you advise somebody about what to do. You will make recommendations and support them with evidence, science knowledge, and explanations.

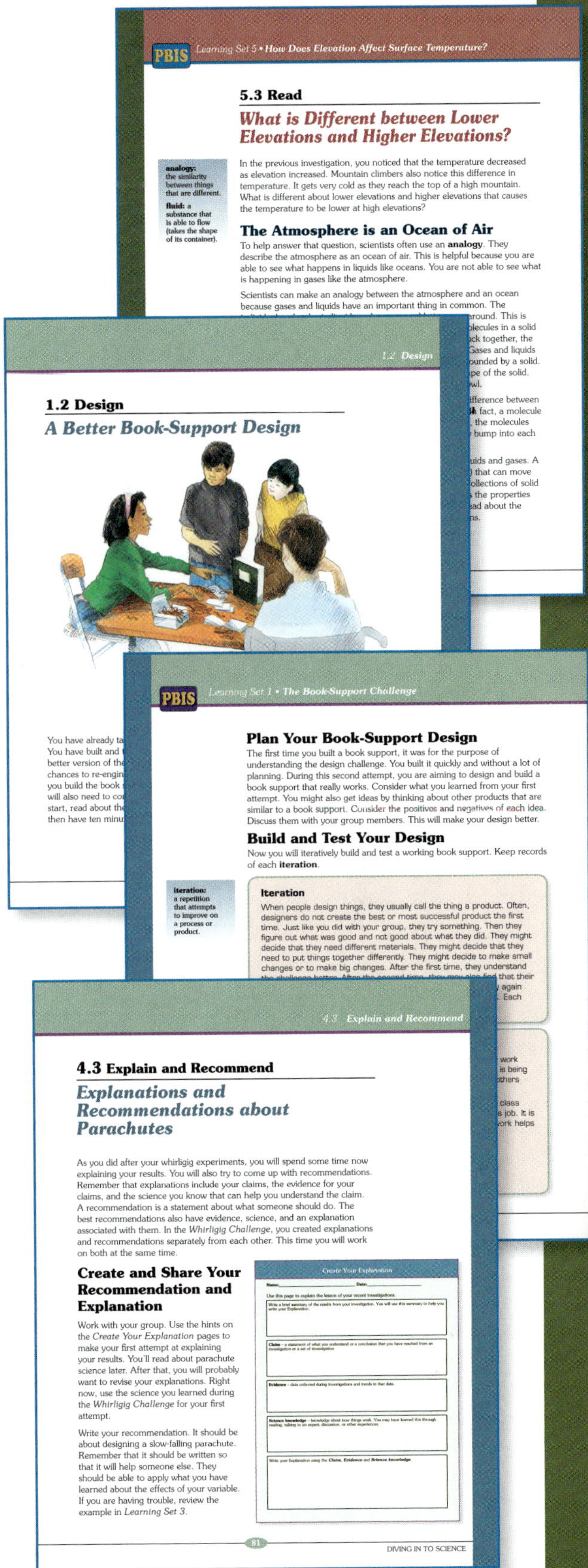

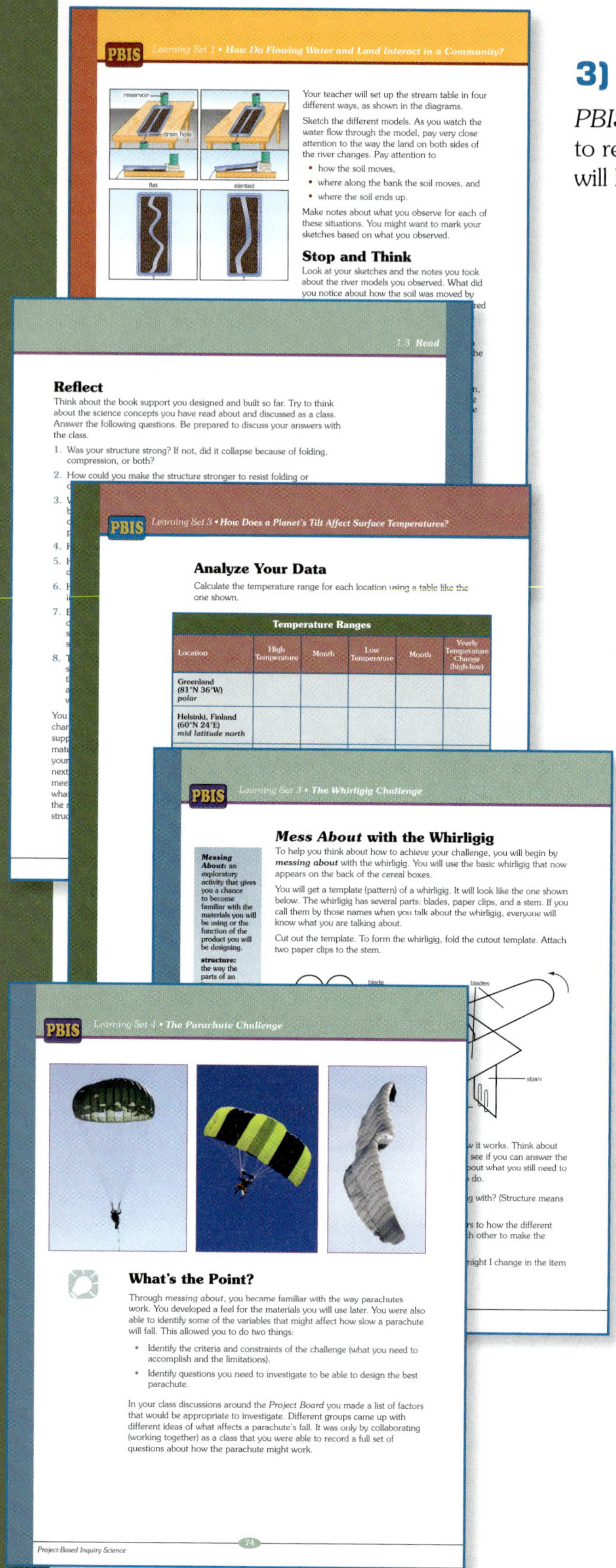

3) Scientists...reflect in many different ways.

PBIS provides guidance to help you think about what you are doing and to recognize what you are learning. Doing this often as you are working will help you be a successful student scientist.

Tools for Making Sense

Stop and Think

Stop and Think sections help you make sense of what you've been doing in the section you are working on. *Stop and Think* sections include a set of questions to help you understand what you've just read or done. Sometimes the questions will remind you of something you need to pay more attention to. Sometimes they will help you connect what you've just read to things you already know. When there is a *Stop and Think* in the text, you will work individually or with a partner to answer the questions, and then the whole class will discuss the answers.

Reflect

Reflect sections help you connect what you've just done with other things you've read or done earlier in the Unit (or in another Unit). When there is a *Reflect* in the text, you will work individually, with a partner or your small group to answer the questions. Then, the whole class will discuss the answers. You may be asked to answer *Reflect* questions for homework.

Analyze Your Data

Whenever you have to analyze data, the text will provide hints about how to do that and what to look for.

Mess About

"Messing about" is a term that comes from design. It means exploring the materials you will be using for designing or building something or examining something that works like what you will be designing. Messing about helps you discover new ideas—and it can be a lot of fun. The text will usually give you ideas about things to notice as you are messing about.

What's the Point?

At the end of each *Learning Set*, you will find a summary, called *What's the Point?*, of the important information from the *Learning Set*. These summaries can help you remember how what you did and learned is connected to the *Big Question or Challenge* you are working on.

4) Scientists...collaborate.

Scientists never do all their work alone. They work with other scientists (collaborate) and share their knowledge. *PBIS* helps you be a student scientist by giving you lots of opportunities for sharing your findings, ideas, and discoveries with others (the way scientists do). You will work together in small groups to investigate, design, explain, and do other things. Sometimes you will work in pairs to figure out things together. You will also have lots of opportunities to share your findings with the rest of your classmates and make sense together of what you are learning.

Investigation Expo

In an *Investigation Expo*, small groups report to the class about an investigation they've done. For each *Investigation Expo*, you will make a poster detailing what you were trying to learn from your investigation, what you did, your data, and your interpretation of your data. The text gives you hints about what to present and what to look for in other groups' presentations. *Investigation Expos* are always followed by discussions about the investigations and about how to do science well. You may also be asked to write a lab report following an investigation.

Plan Briefing/Solution Briefing/Idea Briefing

Briefings are presentations of work in progress. They give you a chance to get advice from your classmates that can help you move forward. During a *Plan Briefing*, you present your plan to the class. It might be a plan for an experiment or a plan for solving a problem or achieving a challenge. During a *Solution Briefing*, you present your solution in progress and ask the class to help you make your solution better. During an *Idea Briefing*, you present your ideas. You get the best advice from your classmates when you present evidence in support of your plan, solution, or idea. Often, you will prepare a poster to help you make your presentation. Briefings are almost always followed by discussions of your investigations and how you will move forward.

Solution Showcase

Solution Showcases usually appear near the end of a Unit. During a *Solution Showcase*, you show your classmates your finished product—either your answer to a question or your solution to a challenge. You also tell the class why you think it is a good answer or solution, what evidence and science you used to get to your solution, and what you tried along the way before getting to your answer or solution. Sometimes a *Solution Showcase* is followed by a competition. It is almost always followed by a discussion comparing and contrasting the different answers and solutions groups have come up with. You may be asked to write a report or paper following a *Solution Showcase*.

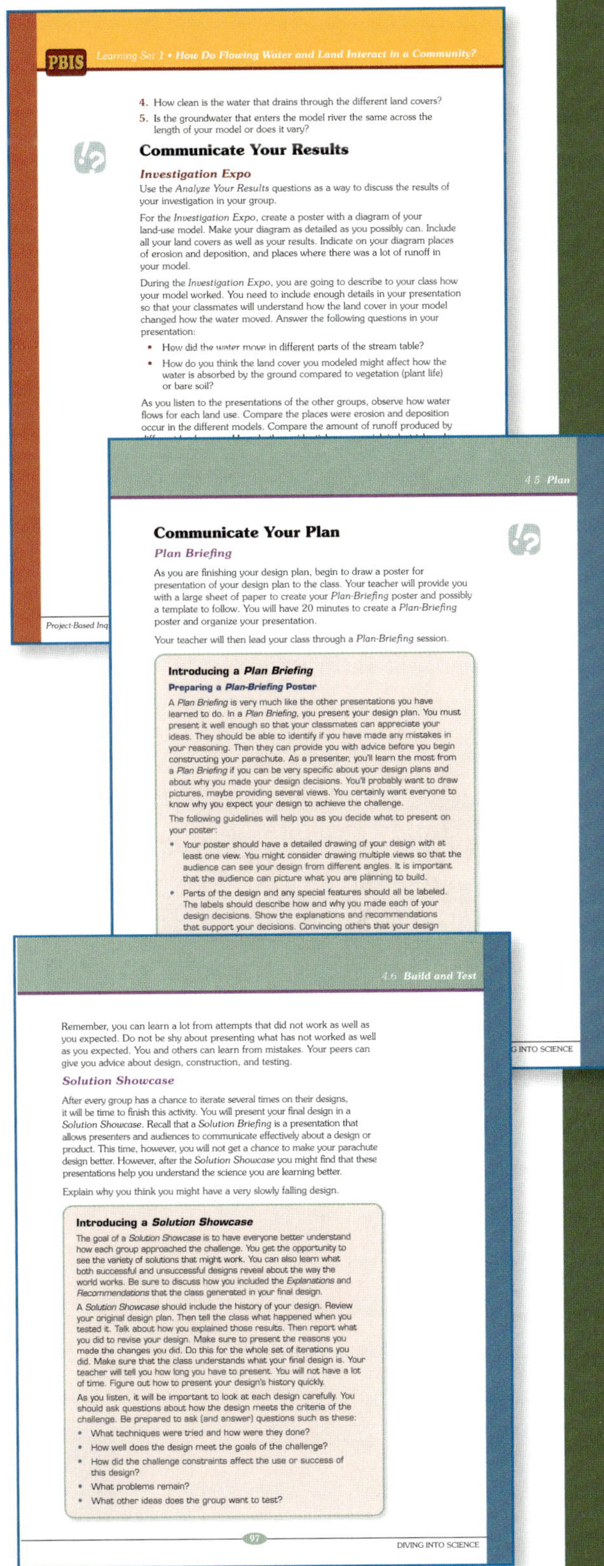
PBIS Learning Set 1 • How Do Flowing Water and Land Interact in a Community?

4. How clean is the water that drains through the different land covers?
5. Is the groundwater that enters the model river the same across the length of your model or does it vary?

Communicate Your Results

Investigation Expo

Use the *Analyze Your Results* questions as a way to discuss the results of your investigation in your group.

For the *Investigation Expo*, create a poster with a diagram of your land-use model. Make your diagram as detailed as you possibly can. Include all your land covers as well as your results. Indicate on your diagram places of erosion and deposition, and places where there was a lot of runoff in your model.

During the *Investigation Expo*, you are going to describe to your class how your model worked. You need to include enough details in your presentation so that your classmates will understand how the land cover in your model changed how the water moved. Answer the following questions in your presentation:

- How did the water move in different parts of the stream table?
- How do you think the land cover you modeled might affect how the water is absorbed by the ground compared to vegetation (plant life) or bare soil?

As you listen to the presentations of the other groups, observe how water flows for each land use. Compare the places were erosion and deposition occur in the different models. Compare the amount of runoff produced by

4.5 Plan

Communicate Your Plan

Plan Briefing

As you are finishing your design plan, begin to draw a poster for presentation of your design plan to the class. Your teacher will provide you with a large sheet of paper to create your *Plan-Briefing* poster and possibly a template to follow. You will have 20 minutes to create a *Plan-Briefing* poster and organize your presentation.

Your teacher will then lead your class through a *Plan-Briefing* session.

Introducing a *Plan Briefing*

Preparing a *Plan-Briefing* Poster

A *Plan Briefing* is very much like the other presentations you have learned to do. In a *Plan Briefing*, you present your design plan. You must present it well enough so that your classmates can appreciate your ideas. They should be able to identify if you have made any mistakes in your reasoning. Then they can provide you with advice before you begin constructing your parachute. As a presenter, you'll learn the most from a *Plan Briefing* if you can be very specific about your design plans and about why you made your design decisions. You'll probably want to draw pictures, maybe providing several views. You certainly want everyone to know why you expect your design to achieve the challenge.

The following guidelines will help you as you decide what to present on your poster:

- Your poster should have a detailed drawing of your design with at least one view. You might consider drawing multiple views so that the audience can see your design from different angles. It is important that the audience can picture what you are planning to build.
- Parts of the design and any special features should all be labeled. The labels should describe how and why you made each of your design decisions. Show the explanations and recommendations that support your decisions. Convincing others that your design

4.6 Build and Test

Remember, you can learn a lot from attempts that did not work as well as you expected. Do not be shy about presenting what has not worked as well as you expected. You and others can learn from mistakes. Your peers can give you advice about design, construction, and testing.

Solution Showcase

After every group has a chance to iterate several times on their designs, it will be time to finish this activity. You will present your final design in a *Solution Showcase*. Recall that a *Solution Briefing* is a presentation that allows presenters and audiences to communicate effectively about a design or product. This time, however, you will not get a chance to make your parachute design better. However, after the *Solution Showcase* you might find that these presentations help you understand the science you are learning better.

Explain why you think you might have a very slowly falling design.

Introducing a *Solution Showcase*

The goal of a *Solution Showcase* is to have everyone better understand how each group approached the challenge. You get the opportunity to see the variety of solutions that might work. You can also learn what both successful and unsuccessful designs reveal about the way the world works. Be sure to discuss how you included the *Explanations* and *Recommendations* that the class generated in your final design.

A *Solution Showcase* should include the history of your design. Review your original design plan. Then tell the class what happened when you tested it. Talk about how you explained those results. Then report what you did to revise your design. Make sure to present the reasons you made the changes you did. Do this for the whole set of iterations you did. Make sure that the class understands what your final design is. Your teacher will tell you how long you have to present. You will not have a lot of time. Figure out how to present your design's history quickly.

As you listen, it will be important to look at each design carefully. You should ask questions about how the design meets the criteria of the challenge. Be prepared to ask (and answer) questions such as these:

- What techniques were tried and how were they done?
- How well does the design meet the goals of the challenge?
- How did the challenge constraints affect the use or success of this design?
- What problems remain?
- What other ideas does the group want to test?

97 DIVING INTO SCIENCE

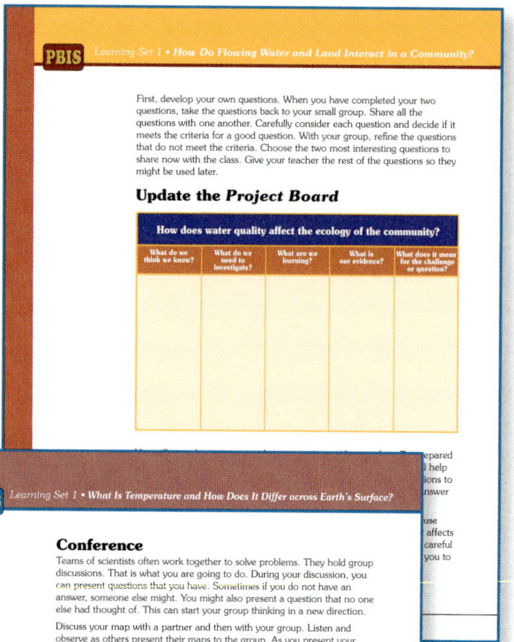

Learning Set 1 • How Do Flowing Water and Land Interact in a Community?

First, develop your own questions. When you have completed your two questions, take the questions back to your small group. Share all the questions with one another. Carefully consider each question and decide if it meets the criteria for a good question. With your group, refine the questions that do not meet the criteria. Choose the two most interesting questions to share now with the class. Give your teacher the rest of the questions so they might be used later.

Update the *Project Board*

How does water quality affect the ecology of the community?				
What do we think we know?	What do we need to investigate?	What are we learning?	What is our evidence?	What does it mean for the challenge or question?

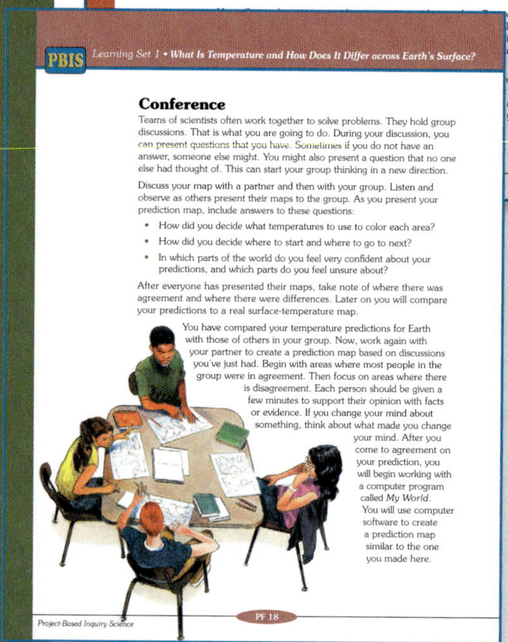

Learning Set 1 • What Is Temperature and How Does It Differ across Earth's Surface?

Conference

Teams of scientists often work together to solve problems. They hold group discussions. That is what you are going to do. During your discussion, you can present questions that you have. Sometimes if you do not have an answer, someone else might. You might also present a question that no one else had thought of. This can start your group thinking in a new direction.

Discuss your map with a partner and then with your group. Listen and observe as others present their maps to the group. As you present your prediction map, include answers to these questions:

- How did you decide what temperatures to use to color each area?
- How did you decide where to start and where to go to next?
- In which parts of the world do you feel very confident about your predictions, and which parts do you feel unsure about?

After everyone has presented their maps, take note of where there was agreement and where there were differences. Later on you will compare your predictions to a real surface-temperature map.

You have compared your temperature predictions for Earth with those of others in your group. Now, work again with your partner to create a prediction map based on discussions you've just had. Begin with areas where most people in the group were in agreement. Then focus on areas where there is disagreement. Each person should be given a few minutes to support their opinion with facts or evidence. If you change your mind about something, think about what made you change your mind. After you come to agreement on your prediction, you will begin working with a computer program called *My World*. You will use computer software to create a prediction map similar to the one you made here.

Project-Based Inquiry Science PF 18

Update the Project Board

Remember that the *Project Board* is designed to help the class keep track of what they are learning and their progress towards a Unit's *Big Question* or *Challenge*. At the beginning of each Unit, the class creates a *Project Board*, and together you record what you think you know about answering the *Big Question* or addressing the *Big Challenge* and what you think you need to investigate further. Near the beginning of each *Learning Set*, the class revisits the *Project Board* and adds new questions and things they think they know. At the end of each *Learning Set*, the class again revisits the *Project Board*. This time you record what you have learned, the evidence you've collected, and recommendations you can make about answering the *Big Question* or achieving the *Big Challenge*.

Conference

A *Conference* is a short discussion between a small group of students before a more formal whole-class discussion. Students might discuss predictions and observations, they might try to explain together, they might consult on what they think they know, and so on. Usually, a *Conference* is followed by a discussion around the *Project Board*. In these small group discussions, everybody gets a chance to participate.

What's the Point?
Review what you have learned in each *Learning Set*.

Stop and Think
Answer questions that help you understand what you've done in a section.

Communicate
Share your ideas and results with your classmates.

Record
Record your data as you gather it.

PBIS

DIGGING IN

As a student scientist, you will...

What's the Big Question?

How Do Scientists Work Together to Solve Problems?

Welcome to your new science Unit. This Unit and other *PBIS* Units you complete this year will offer you exciting challenges and opportunities to learn science. Science involves learning very interesting facts, but that's not all science is. A large part of science is the ability to analyze and make sense of the world around you in an organized and logical way. Scientists learn how to do this to be successful at what they do. But scientists are not the only ones who can benefit from this kind of reasoning; you may also find it useful.

In this Unit, you will learn how to tackle problems and challenges as a scientist or engineer does. There is a lot scientists do to make sure that they ask questions and investigate answers in an organized and logical way. Engineers also need to solve problems in an organized and logical way. You will experience and use many of the practices that scientists and engineers use. You are not expected to learn everything about being a scientist or engineer in just a few weeks, but you will learn a lot. The lessons you learn will help you be successful in science class this year. They will also help you in future science classes and even in your life!

Your *Big Question* in this Unit is to understand how scientists solve problems. To help you answer the *Big Question*, you will work on three challenges in this Unit. Each one will give you a chance to learn a few practices and behaviors. Then you will use what you have learned to answer the *Big Question*: *How do scientists work together to solve problems?*

Some of you may have already begun learning about what scientists and engineers do and how they work together. For you, much of this Unit will be review. That review will be useful to you. It will give the members of your class a chance to learn to work together. It will also allow you to share what you have learned in other years with your new classmates. You will also find new things to learn in this Unit. This Unit introduces new content and practices of scientists and engineers that you may not have discussed before.

Have fun being student scientists!

Learning Set 1

The Build a Boat Challenge

Imagine that you and your friends are visiting an historic fort in a nearby state park. It is getting dark, and the park and fort will be closing soon. You and four friends walk over the drawbridge just as the gatekeeper is locking the door and preparing to raise the bridge. When you get to the parking lot, you realize that two of your friends are missing!

You rush back to the gate, but it is locked, and the drawbridge has been raised. There is a stream running into the fort through a very small opening in the wall. The opening, which is only about 5 cm wide, is the only way in.

The gatekeeper is still there, so you explain to him that two of your friends are locked inside the fort. "Oh, dear," says the gatekeeper. "We can't get back in until tomorrow morning. Once I lock up, the gates cannot be opened again until morning. If we try to unlock them from the outside, the alarm will go off. The only way to open the gates without setting off the alarm is from the inside, and it takes six different keys! I am afraid your friends are stuck overnight."

"What about the stream and the opening in the wall?" you ask. "I have an idea. What if we build a little boat and float the keys down the stream and in through the opening in the wall? But, what can we use to build the boat?"

You remember that there are some sandwiches wrapped in aluminum foil still in your backpack. You and your friends decide to build a boat out of foil to see if you can get it to hold the weight of the keys.

"Hold on there!" says the gatekeeper. "If the keys fall into the stream, we won't be able to open the gates at all. You will need to make sure that your boat can float for at least 20 seconds so the keys will make it all the way into the fort."

Your group decides to try. The gatekeeper lets you use the keys and a bucket of water to determine if your plan will work. One problem you have is that the supply of foil is limited, so you can use only a 5" × 5" square of foil. The other problem is that the gatekeeper tells you that he has to leave in 20 minutes!

You begin by discussing the problem with your group, and then you will try to design a boat that can carry six keys downstream for at least 20 seconds. Your group will have 10 minutes to complete the boat.

1.1 Understand the Challenge

Identify Criteria and Constraints

criteria (singular: criterion): goals that must be satisfied to successfully achieve a challenge.

constraints: factors that limit how you can solve a problem.

Before you start, it is a good idea to make sure you understand what your challenge is. Design challenges have two parts: criteria and constraints.

Criteria are things that must be satisfied to achieve the challenge. For the boat challenge, this will include the job the boat must do and the specific details of how the boat will do that job. To get the job done, you may have thought about buying a boat. However, the challenge does not offer that as a possibility. So, one **criterion** is that you must build the boat.

Constraints are factors that limit how you can solve a problem. Think about the constraints that have been placed on you for this challenge. You can use only a 5" × 5" square of aluminum foil. This makes the size of the foil a constraint. Think about other constraints that have been placed on you for this challenge.

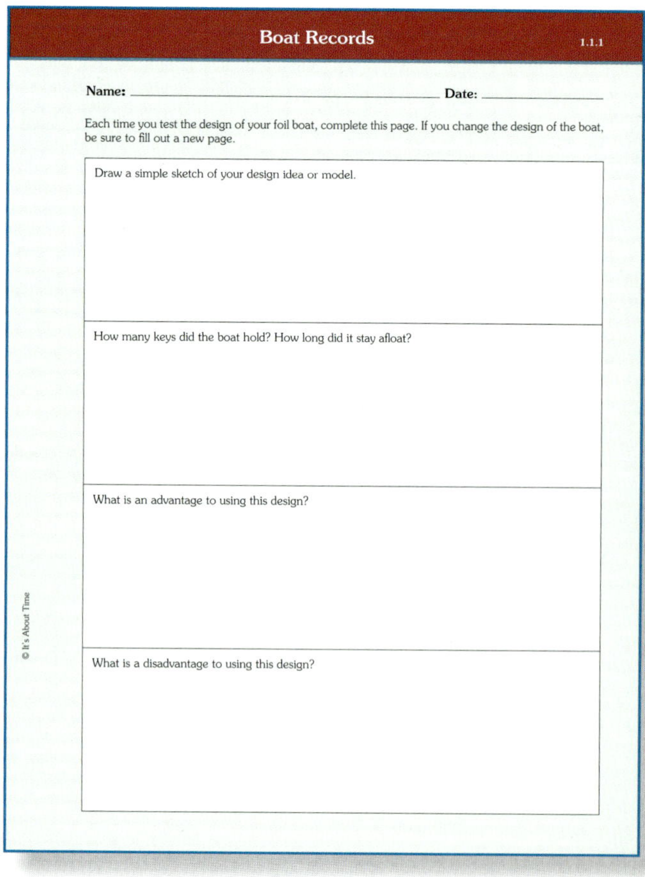

Boat Records 1.1.1

Name: ______ Date: ______

Each time you test the design of your foil boat, complete this page. If you change the design of the boat, be sure to fill out a new page.

Draw a simple sketch of your design idea or model.

How many keys did the boat hold? How long did it stay afloat?

What is an advantage to using this design?

What is a disadvantage to using this design?

© It's About Time

Build and Test Your First Boat

To help you think about how to achieve your challenge, you will begin by drawing a sketch of what your boat will look like. As you sketch your boat, think about how the product you are designing is supposed to function, or work. Remember that you have only a 5" × 5" square of foil to work with, and the boat you build must be able to carry six keys for at least 20 seconds. After everybody has had a chance to think and sketch, spend some time sharing your ideas with your group members. Then work together to build and test your boat. Try out different ideas. Think about which ones seem to work better. You will have a total of 10 minutes.

Recording Your Results

Keep track of the designs you attempt for this test boat. Record your sketches on a *Boat Records* page. All group members should keep a record of the final boat design on their own pages.

Communicate Your Work

Share Your Designs

It is time to share your design with your classmates. Groups should take turns presenting their boats to the rest of the class. After each presentation, the teacher will test your design to see if it meets the criteria. There will be time for classmates to ask questions of the presenting group.

As you present your boat, try to answer these questions:

- How is your design constructed?
- Why did you design it the way you did?
- How did the challenge constraints affect the design?
- What things did you think about and try before getting to this design?

As you listen to everybody's reports, make sure you understand the answers to these questions for each report. If you do not think you have heard answers to each question, ask questions (like those above). Be careful to ask your questions respectfully.

After each boat is tested, the class should quickly discuss and agree about how well the design fulfills the challenge's criteria.

Materials

- **5" × 5" square of foil**
- **6 keys**
- **stream table**

Reflect

Answer each of the following questions. Draw and label a sketch of your group's boat, showing what its good qualities are, what problems it has, and what you might want to change.

1. Which criteria did your design meet?
2. What qualities make your boat a good design?
3. What are the problems with your current design?
4. What did you learn from other groups that will help you with another design if you have another chance?

Update Your Criteria and Constraints

Now that you have tried achieving the challenge, you have found that there is more to think about than you imagined earlier. You may now realize that the criteria and constraints are different than you had first expected. For example, you know that six keys weigh quite a bit. You will soon have a chance to design and build a better boat. Before that, review your list of criteria and constraints. Update the list, making it more accurate. A more accurate list will help you design a better-performing boat.

What's the Point?

You now understand what is required for achieving this challenge better than you did when it was first presented to you. People often try to solve a problem without taking the time to think about it first. If you do not understand a problem well, your solution will not be the best one. In fact, you might fail. Each time you are presented with a new problem, take the time to think. Identify what you have to achieve (criteria). Also, consider what limits you are working under (constraints). You may also find it useful to explore the materials you will be using. Making a first, simple try at a solution may give you a better understanding of the problem you have to solve. With better understanding of the problem, and what is required to solve it, you are more likely to be successful.

1.2 Design

Build a Better Boat I

You have just completed your first attempt at building an aluminum-foil boat. You also talked about the design ideas and products every group created. Along the way, you learned about some new ideas that work well, as well as some ideas that do not work so well. Many of you might like to have another try at building a better boat. You are going to have that chance. In this science class, you will have many chances to modify solutions to problems or challenges. Each try you make is called an **iteration**. Before you start your next iteration of the Boat Challenge, read the rest of this page and the next page to prepare for what's ahead. You will then have 10 minutes to design and build a better boat.

iteration: a repetition that attempts to improve on a process or product.

Plan Your Boat Design

You built your first boat quickly and without a lot of planning. It may or may not have worked well. During this second attempt, you have a chance to design and build a boat that really works. Consider what you learned from your first attempt. Did it meet the criteria of the challenge? If not, what can you do to improve the design so it does meet the criteria? You may get ideas by thinking about the different designs that your classmates came up with. Think about the designs that worked well and those that did not. Discuss these ideas with your group members. This will make your design better.

Be a Scientist

When people design things, they usually call it a product. Often, engineers do not create the best or most successful product the first time. Just like you did with your group, they try something. Then they figure out the strengths (what was good), and the weaknesses (what was not good) in the design. They might decide that they need different materials. They might decide that they need to put things together differently. They might decide to make small changes or to make big changes. After the first time, they understand the challenge better. After the second time, they may also find that their solution is not as good as they would like. Engineers often try again and again before they get the product just the way they want it. Each try is called an iteration.

Build and Test Your Design

Build and test a working boat that can carry six keys for at least 20 seconds. Keep records of each iteration.

Recording Your Work

Materials
- **5" × 5" square of foil**
- **6 keys**
- **stream table**

Keep track of the number of keys your boat can keep afloat and the amount of time it can stay afloat. Record this information on new *Boat Records* pages. Each of you should fill in your own page.

One part of the *Boat Records* page has room for you to draw your design after you have finished building it. Each time you create a new design, continue to fill out the page. This way you will have a record of all the designs you attempted. You will be able to see how you made improvements along the way.

After completing your boat, you will present it to the rest of the class. During your presentation, have your *Boat Records* pages handy to report your results to the class.

The designers of this boat went through many iterations to improve the boat's design.

Be a Scientist

Keeping Records

Scientists and engineers always record their work as they go. To record means to write, illustrate, or diagram what is being done. This important step allows scientists and engineers to accurately report their findings to others and helps them design future investigations or products.

You probably had to keep records of your work in class before. Keeping records is important for scientists and engineers, and it is also important for you as a student scientist or engineers. Recording your work helps you to do the following:

- Share your work with others.
- Remember what you did and decided along the way.
- Remember why you decided to do those things.
- Make decisions about what investigations to do next.

Communicate Your Results

Solution Briefing

After you designed and built your first boat, you presented it to the class. You will also present your redesigned boat to the class. This time you will present more formally in a *Solution Briefing*. In a *Solution Briefing*, you present your solution in a way that will allow others to evaluate how well it achieves criteria and to make suggestions about how you might improve it. Before you start preparing, read more about *Solution Briefings* on page 13.

Get ready for this briefing by preparing a presentation that answers these questions:

- How is your design constructed?
- What materials did you use?
- Why did you design and build it the way you did?
- How does the design meet the criteria?
- How did the constraints affect the design?

- What past experiences helped you make your design?
- What problems remain?
- What things did you try along the way?
- How well does your boat work? What else do you want to test?

As you listen to your classmates' presentations, make sure you understand the answers to the questions on the previous page. If you do not understand something, or if they did not present something clearly enough, ask questions. You can use the questions on the previous page as a guide. When you think something can be improved, make sure to contribute your ideas. Be careful to ask your questions and make your suggestions respectfully. Record the interesting things you are hearing on your *Solution-Briefing Notes* page.

Solution-Briefing Notes 1.2.1

Name: ________________ Date: ________

Design Iteration: ________________

Design or group	How well it works	What I learned and useful ideas		
		Design ideas	Construction ideas	Science ideas

Plans for our next iteration

Be a Scientist

Introducing a *Solution Briefing*

A *Solution Briefing* is useful when you have made one or more attempts to solve a problem or achieve a challenge and need some advice. It gives you a chance to share what you have tried and learned. It also provides an opportunity for you to learn from others. You can ask advice of others about difficulties you are having.

Real-life engineers present their designs to each other and to others several times as they work on design projects. A team of engineers sets up their design or design plans, and everybody gathers around. They make sure everyone can see. The design team presents their design plan to everyone. The other engineers ask questions and give helpful advice about ways to improve the design.

You will do the same thing. In a *Solution Briefing*, each team presents their solution for others to see. Then teams take turns presenting to the class. Other classmates ask questions and offer helpful advice. You might walk around the class from design to design, or teams might take turns presenting in front of the class.

A *Solution Briefing* works best when everyone communicates well. Before you present your design to the rest of the class, think about what might be important to share. What aspects of your design should you present? What parts do you want to discuss with others? You need to be ready to justify to others what you decided to do and why.

When you are listening to a *Solution Briefing*, it is important to pay close attention. Look at each design or plan. Think about questions you would like to ask about the design.

Each time you hold a briefing, you will take notes. You will fill out a *Solution-Briefing Notes* page as you listen to each group's presentation.

Collaboration is a group effort. A team of designers share sketches and ideas at the innovation and design firm, IDEO.

Reflect

Following your *Solution Briefing*, answer the following questions. Discuss your answers and how they may help you better achieve the *Boat Challenge*.

You may find looking at your drawings and your *Solution-Briefing Notes* page helpful as you answer the following questions. Be prepared to discuss your answers with your class.

1. Before each boat design was tested, what did you think would happen when the keys were placed in it?
2. Which boats worked the way you thought they would? Which worked differently than you expected? For the ones that worked differently, what might help you to understand why?
3. What qualities make your boat a good design?
4. What are the problems with your current design?
5. What can you borrow from other designs to make yours work better?
6. What do you need to learn more about to make a better design?

Be a Scientist

Build on and Benefit from One Another's Ideas

You ask questions and offer suggestions during a *Solution Briefing*. When you do this, you are **collaborating** with one another. You are working together. You offer your ideas for others to think about. You provide suggestions that might help them improve their solutions. Sometimes you learn something that you want to try yourself.

Other teams may come up with solutions or ideas that you want to borrow and make better. You may also find that other teams have used your suggestions. Is the other team copying from you? Are you copying from them?

Think about the other team's success as your success when they use something you suggested. Help them see that your success is theirs if you borrowed something from them.

collaborate: to work together.

When movie actors receive awards, they often thank many people. Even though they are getting the award, they know that it takes lots of people to put a movie together. It is the same with scientists and engineers.

They would not be able to solve problems or learn new things without building on the work of others. Scientists and engineers write papers, or articles, in journals. They tell others what they have discovered. Others read those papers, talk about the ideas, and ask questions. When someone improves on an idea, they write a paper about it and publish it for others to read. Other people can then improve on their ideas. Each time someone publishes a paper, they give credit to the scientists or engineers they got their ideas from. They do this by **citing** the paper that the idea was presented in. This is the way science is done.

Engineers and inventors submit patents, telling others how to make something work. When other engineers and product designers use these ideas in their own products, they give credit and pay **royalties** for using those ideas. This is how product design and invention are done.

cite: to quote something or somebody.

royalty: in this case, a fee paid to the owner of a patent for permission to use it.

These students, like a group of scientists, are collaborating on their design project. They are all contributing questions and information. This sharing of ideas and building on the past work of other scientists are the ways the best science research is accomplished.

Be a Scientist

Copying versus Crediting

When you build on someone else's idea, it is important to give them credit. Why isn't this "copying?" Copying means taking the work of someone else and claiming it as your own. If you simply build what some other group built, that is copying. But if you add to another group's idea and acknowledge from where you got your idea, you are doing what scientists and engineers do. When you explain how you used their ideas and made them better, you are adding your contribution to theirs.

This means that you will have to keep good records of where you got your ideas. When you use someone else's ideas, always record from whom you borrowed the idea. Record how you included it in your design and why you did it that way. Then, make sure to give credit to the other person or group in your presentations.

What's the Point?

The boats you built the second time were probably more successful than the first ones. In general, the more chances you get to iterate, the better your solution will turn out. Each attempt you make is an iteration. Each time you make another attempt, you can do better because you use knowledge gained from the previous attempt. You also identify more that you need to learn each time you try to achieve a challenge. Answering the questions that come up before trying again gives you a chance to do even better. This iterative approach to design and problem solving is what scientists do. Use this approach whenever you have a problem to solve.

You read that there is a difference between copying and building on the ideas of others. As you took part in the first *Solution Briefing*, you may have seen some design ideas that worked well. You may have used some of those ideas to improve your boat. Others may have used some of your ideas. When you claim someone else's ideas as your own, it is copying. If, however, you give credit to a group for their idea, you are building on the work of others. This is how scientists work and how science grows. Science builds on the ideas of others.

You probably have begun to realize the importance of keeping records as you work on a design challenge. You used your records when you presented your ideas during the *Solution Briefing*. You also got a chance to see other solution ideas during the briefing. You saw what works and what does not work as well. This may help you develop better ideas as you continue to try to achieve your challenge. You can learn a lot from attempts that "failed" as well as ones that succeeded. In either case, the goal is to understand the challenge better and create better solutions.

1.3 Read

The Science of Boat Design

matter: anything that has mass and takes up space.

density: the amount of matter in a given amount of space.

buoyant force: the upward push that keeps objects floating in liquid.

volume: the amount of space that something takes up.

atom: a small particle of matter.

molecule: the combination of two or more atoms.

You have just finished your first attempts at building an aluminum foil boat. You also talked about the design ideas and products of other groups. You discovered some ideas that worked well and others that did not. You identified some questions that you want answered before you try again. Soon you will have another chance to build a better boat. Before you do, you will read and think about the science concepts that explain how boats work, and you can then apply this knowledge to your next boat design. To understand what makes things float, it is important to learn about three science concepts—**matter**, **density**, and **buoyant force**. They are all important to making your foil boat carry more keys.

Matter

All objects of any form (solid, liquid, or gas) are made up of matter. All matter has mass and takes up space. The amount of space that something takes up is its **volume**. The boat you are trying to build is made up of matter, and so is the water the boat floats on. Matter is made of extremely small particles called **atoms**. These atoms combine with other atoms to form larger particles called **molecules**. Molecules attach to one another to form all the objects that you see, touch, hear, taste, and smell.

Density

One factor that affects whether or not something can float is its density. Density is the scientific word for the amount of matter in a certain amount of space. It is a measure of how tightly the molecules making up matter are packed together in the space. The more room the molecules have in a given space, the less dense the matter will be.

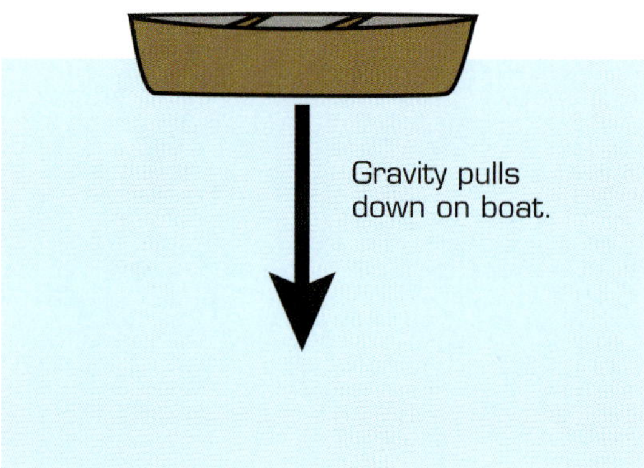

If you have a cardboard box full of plastic bubble wrap, it will be lighter than the same-sized box full of books. The different materials in each box take up the same amount of space, but each contains a different amount of matter. Since a book-filled box has more matter than a box filled with bubble wrap, the box of books has greater density.

We think of books as being heavier than bubble wrap, but that is misleading. What we are really thinking about in that case is the density of the materials. A book taking up a certain amount of space will be heavier than a piece of bubble wrap taking up the same amount of space. This means that books are denser than bubble wrap. There is more matter in a book than in a section of bubble wrap of the same size as the book. That is why a box of books will be heavier than the same box filled with bubble wrap. For the same volume, the more dense material will be heavier than the less dense material.

Molecules that make up the matter in a book are tightly packed together and do not have much space between them. In the bubble wrap, the molecules have a lot of space between them because each bubble contains a lot of air—air is a gas and is much less dense than a solid. Books are solids and contain a lot less air. This makes bubble wrap less dense than books.

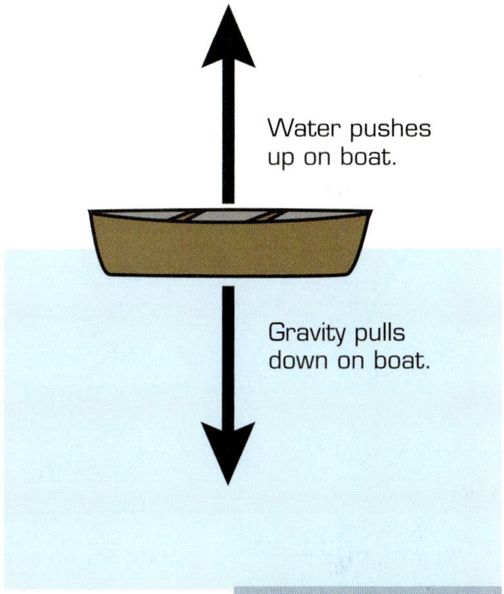

Buoyant Force

A force is a push or pull on matter. The upward push that keeps an object floating is called buoyant force. To understand the buoyant force that makes things float, you first have to understand gravity. You already know a lot about **gravity**. You see and feel the effects of gravity everywhere every day. Gravity is the force that holds you, and all objects, on Earth. It is a force, or pull, between any two objects. All objects have this pull toward other objects. The pull between most objects is small, and unless an object has a lot of mass, you do not feel its pull.

gravity: a pull between two objects. Gravity is the force that holds all objects on Earth.

When one (or both) of the objects is very massive (which means it has a lot of mass and, therefore, has a lot of matter), you can experience gravity's effects. Earth is very massive, and gravity is the force that pulls everything down toward the center of Earth. Because of gravity, almost everything—people, furniture, trains, and dogs—stay put on top of Earth's surface. In your activity with boats, Earth's gravity pulls on the water and keeps the water in the bucket in which you are floating your boats. Gravity also pulls down on the foil boat.

In designing a boat, an important consideration is why some boats stay afloat, while others do not, and sink. This is a question of how much buoyant force the boat produces.

Gravity pulls things toward the center of Earth, but objects do not continue falling toward the center of Earth. The ground, or other surfaces, resists Earth's pull. In the boat-building challenge, the molecules of the water push up on the molecules in the foil boat at the same time that gravity pulls down on the boat. If the buoyant force of the water pushing up on the boat is as strong as the force of gravity pulling the boat down, the boat will float.

The force pulling the boat down is gravity and the water's buoyant force is the upward push helping to keep the boat afloat.

You may have thought that heavy objects sink and light objects float. But some of you might have gotten the heavy keys to float by shaping the boat in different ways. That shows that weight is not the only factor determining if objects float or sink. To illustrate what is happening in the water, look at the way gravity pulls on something that is not in water.

Crumple one of the 5-inch squares of aluminum into a ball, squeezing out as much air as you can. If you place it on a tabletop, you can see that all of the mass of the foil is pushing down on a very small part of the table. Set another 5-inch square of foil flat on the table, and the same mass of foil is now pushing down on a much larger area. The flat piece of foil touches more of the surface of the table. The piece crumpled into a ball touches less of the table's surface. The mass of the foil ball is concentrated into a smaller area of the table, and fewer molecules that make up the table can push back on it.

These children are able to float in the water because the gravity (downward push) of their bodies is equal to the buoyant force (upward push) of the water. Since a flotation device is less dense than the child, it causes a decrease in the overall density of the person wearing it. This means that less of an upward push by the water is needed to keep the wearer afloat.

If you were to place the foil sheet and the foil ball in a bucket of water, what do you think would happen to each? The flat sheet would float. The ball would sink (if *all* the air in the ball had been squeezed out of it). The foil ball would sink because the small area of water in contact with the full mass of the foil does not put enough buoyant force on the foil to keep it above the water. Instead, the water molecules simply slide around and over the foil ball, and it sinks.

When the foil is spread out flat, more of its surface has contact with the water. The same amount of mass from the foil pushes down on a much larger area of water. This creates a situation in which more molecules that make up the water can push up on the foil. As long as the force of gravity pushing down from the foil is equal to the buoyant force pushing up from the water, the foil will float. The flat piece of foil is better able to float because more molecules of water can apply their upward buoyant force to push up on the foil.

The air that fills the open parts in the boat decreases the boat's overall density, making it possible for the buoyant force (upward push) of the water to keep it afloat.

Density and Buoyancy Force

Buoyant force and density work together to affect whether or not something will float. When a boat sits in water, it pushes some of the water away, or displaces it. The water that was pushed away has a certain density. If the boat, including the air in it, is less dense than the water it pushes away, the boat will float. If it is denser than the water it pushes away, the boat will sink. As the density of any object increases, it sinks lower into the water, always displacing an amount of water equal to its weight. The weight of the water that is pushed away, or **displaced**, by the boat is equal to the weight of the boat.

This is a complicated idea to think about. You will have more opportunities to investigate the effects of gravity, buoyant force, and density. For now, think about your challenge. You are trying to figure out how to make the weight of the keys spread out over a large enough area of the surface of the foil boat so that the buoyant force of the water can keep it afloat. Even as you add more keys, the boat will stay afloat as long as you can find ways to spread the weight of the boat and keys over a larger space.

displace: to take the place of.

Reflect

You are going to get another chance to design a boat. You will use the same materials. Think about how your group could design your next boat to better meet the challenge by considering what you now know about gravity, density, and buoyant force. Answering the following questions should help.

1. Think about some of the boat designs that held the most keys. What decisions did the students who designed these boats make that improved the buoyant force of the boat?
2. Did your boat float? If it did not float, why do you think it sank? Discuss *buoyant force* and *density* in your answer.
3. How could you make the boat better able to stay afloat? Remember that you have to float six keys. Use what you have learned about gravity, buoyant force, and density to answer this. Also, take advantage of what you can learn from other groups' designs.

What's the Point?

All objects are made of matter, which is made up of atoms. Atoms combine with other atoms, becoming molecules. All matter, including water and air, is made up of atoms and molecules. Density is the amount of matter in a given amount of space. It is a measure of how tightly the matter is packed together in the space.

For something to float, the force (gravity) pulling down on it cannot be greater than the force (buoyant force) pushing up from the water. To increase the buoyant force pushing up on a boat, you can spread the mass of the object over a greater area of water. This is similar to placing both the flat piece of foil and the crumpled ball of foil in a bucket of water. The flat piece floats, while the crumpled piece, if all the air has been removed, will sink.

When the foil is spread out, more of its surface has contact with the water. The same amount of mass from the foil pushes down on a much larger area of water. More molecules that make up the water can push up on the foil, or any size boat. The greater the surface of an object that touches the water, the more molecules of water can apply their buoyant force and push up the object.

1.4 Design

Build a Better Boat II

You now have the knowledge and experience to design a boat that will hold six keys. However, just when you thought you had it all figured out, the gatekeeper tells you that your friends will need two more keys to turn off the alarms. He wants you to show him a final design that will now carry eight keys and an explanation of why you think this will work. He will then be confident in giving you the keys to the fort.

You will discuss as a class what you think will be the best design to support the additional weight of two more keys long enough to float them into the fort. As you discuss the boat design, keep in mind what you just read about density and buoyancy. Use this knowledge to help you design the right boat.

Update Your Criteria and Constraints

Now that the challenge has changed, review the list of criteria and constraints. Update these lists. Then consider these changes as you design and build your new boat.

Materials

- **5" × 5" square of foil**
- **8 keys**
- **stream table**

Plan, Build, and Test Your Design

As you design your new boat, you are welcome to use ideas other groups have developed, but you must make sure to give them credit. Once again, be sure to keep track of the design and the number of keys the boat holds. Record your results on a *Boat Records* page.

After completing your boat, your group will present your new solution to the rest of the class. During your presentation, have your *Boat Records* pages handy to report your results to the class. Good luck!

Communicate Your Solution

Solution Briefing

Now it is time to share your new boat with the class. Once again, you will participate in a *Solution Briefing*.

As before, spend some time preparing for your presentation. Be prepared to answer questions such as the following:

- How is your design constructed?
- What materials did you use, and how much of each?
- Why did you build it the way you did?
- How does the design meet the criteria?
- How did the constraints affect the design?
- What past experiences helped you make your design?
- What science knowledge helped you make your design?
- What problems remain?
- What else do you want to test?

When you present your boat design, your group will need to justify the design decisions you made and share the results of any other designs you created. As before, keep notes on a *Solution-Briefing Notes* page. As you listen to the presentations, remember to ask questions if anything is unclear.

Reflect

Answer the following questions. Be prepared to discuss your answers with your class.

1. Review the criteria and constraints for the first *Boat Challenge* and then for the second *Boat Challenge*. Which criteria and constraints are different in the second challenge?
2. What changes did you make in your design to address the new criteria and constraints?
3. How did you change your original design to include the science of density and buoyancy?
4. What criteria and/or constraints were you unable to meet? Why?

What's the Point?

Now that you have designed your boat two times, you have seen how useful iteration is. Each time you iterated on your design, you had a chance to use what you learned from the last time. Each time, as a result of using new knowledge, you made your boat better.

Sometimes the new ideas you had were based on new science you learned, and sometimes you learned from what other groups had done. You may have remembered experiences that helped you form ideas. Ideas can come from all of these places. It is important, when ideas are borrowed from others, to give them credit. This is how the fields of science and engineering make progress. Also, people feel good when others use their ideas and give them credit.

Learning Set 1

Back to the Big Question

How do scientists work together to solve problems?

Over the past few days, you and your classmates have been working to create a boat that can carry six or eight keys for at least 20 seconds. The last boat you built was probably a lot better than the first one. During this activity, you took part in several practices that scientists use when they solve problems. Think about some of the things you did in this *Learning Set*.

You identified the criteria and constraints of your challenge. Criteria are the requirements your solution must meet. Constraints are the factors that put limits on your solution. You also saw how criteria and constraints could change as you attempt to solve the problem.

You learned that there is a difference between copying and building on the ideas of others. You saw some designs of other groups that may have looked very good. In your next attempt, you may have used some of these ideas. Others might have used some of your ideas. This is how scientists work and how science grows as a field. Science builds on the ideas of others.

Scientists work together. They support each other. Working together to build ideas and understanding is called collaboration. In this class, you will collaborate to solve problems or meet challenges. As you collaborate, you will share ideas with others. Others will share ideas with you. One way you collaborated was to participate in a *Solution Briefing*. Scientists often present solutions or ideas while they are trying to solve problems.

Iteration can help you achieve a challenge or solve a problem. You probably saw that it is not always easy to achieve success the first time you try something. But once you shared and saw the ideas of others and learned some scientific concepts, you were able to plan and build a better boat. Scientists also use iteration when solving problems.

In this *Learning Set*, you probably approached problem solving differently than you have in the past. You now have a better idea of how scientists work together to solve problems.

Reflect

Define or describe the following ideas and why they are important to working together as student scientists. Be prepared to discuss your answers with your class.

- criteria and constraints
- iteration
- collaborating
- building on the ideas of others
- record keeping
- asking questions
- using science knowledge

Learning Set 2

The Lava Flow Challenge

Nothing shows the power of nature quite like a volcanic eruption. Streams of red-hot lava flowing over land make spectacular photographs. However, these are life-threatening situations. In this *Learning Set*, you will be working as part of a team of scientists. You have been hired by a company that makes measurement equipment. The company wants you to develop an accurate procedure for measuring lava flow. You will have to send the company a procedure along with evidence that your procedure is accurate.

The Hawaiian Islands are volcanic islands. They are made of volcanic material laid down from repeated eruptions. Some of these volcanoes are still active today. Towns near volcanoes usually develop emergency evacuation plans that allow them to move people out of the area safely and quickly if a volcano begins to erupt. Towns usually develop one plan for quick evacuation and other plans for slower evacuations. When a volcano begins to erupt, towns choose the right plan based on how fast the lava is flowing. To determine this, the people of the town need an accurate procedure for measuring the lava flow.

A **volcano** is a place on Earth from which melted **rock**, ash, gases, and other materials can escape from beneath Earth's surface. Volcano is also the name for the mountain created by the hardened rock. Rock, melted deep in Earth, is called **magma**. When the magma flows out of the volcano, it is known as **lava**. Lava and gases escape through an opening on Earth's surface called a **vent**.

When lava is very thick, it moves slowly. However, thick lava is likely to cause explosions because hot gases get trapped in it. When lava is thinner, it moves much more quickly. Thinner lava is more likely to trap people who live in the surrounding area. It is important to figure out a way to accurately measure the rate of lava flow so that townspeople will know how much time they have to evacuate during an eruption.

volcano: a place on Earth from which melted rock, ash, gases, and other material can escape from beneath Earth's surface.

rock: a naturally formed, non-living solid mass composed of grains of Earth material.

magma: melted rock found beneath Earth's surface.

lava: melted rock that has reached the surface of Earth.

vent: an opening on Earth's surface that allows lava, ash, gases, or other volcanic material to escape.

You will work with a partner on this challenge. You will be given materials to imitate the flow of lava over the land. You will use a plastic plate, model lava (dish soap), and a stopwatch. You and your partner will develop a procedure for measuring how fast the dish soap moves across the plate. Imagine that this is the same way lava would flow over land. You will share your procedure with your class, and together, the class will develop a procedure that everyone agrees is accurate and reliable.

2.1 Understand the Challenge

Identify Criteria and Constraints

Before you get started, make sure that you understand what your challenge is. There are two features of the challenge: the criteria and the constraints.

Remember that criteria are things that must be satisfied to achieve a goal or answer a question. Constraints are factors that will limit how you can go about doing that. Think about and record the goals of the challenge. Think about the limits that have been placed upon you for this challenge. For example, you cannot work with actual lava.

What's the Point?

You have been given a new challenge. Remember, to be successful, you need to understand the parts of the challenge. You need to figure out what you need to achieve (criteria). You must also consider the limits you are working under (constraints). By identifying the criteria and constraints, you are more likely to be successful with your challenge.

The largest of the chain of Hawaiian Islands, the big island of Hawai'i consists of five volcanoes. Several of the volcanoes have erupted over the past 200 years. During some of the spectacular eruptions, lava as hot as 1204°C (2200°F) flows out of the volcanoes.

2.2 Investigate

Modeling Lava Flow I

Because you cannot use actual lava, you are going to make a **model** of lava flow. You will be using dish soap and a plastic plate to **simulate** lava flowing across a landscape. You might say that this model is not realistic because dish soap is not as hot as lava. That is true, but your model is not investigating the temperature of lava. You are looking at how fast lava flows, and dish soap flows very much like some types of lava. This makes the soap and plastic plate a good set of materials for modeling lava flow.

Models and simulations help scientists learn. The simulations you are doing will help you learn how to measure the rate at which lava flows. You will work with several other models and simulations in this Unit and throughout PBIS.

model: a representation of something in the world.

simulate: use a model to imitate or act out real-life situations.

Be a Scientist

Using Models and Simulations

A model is a representation of something in the world. One model that you know is a globe. The parts of the globe represent parts of Earth. Scientists use models to investigate things that are too difficult or too dangerous to examine in real life. The models they use are at a size that people can easily examine.

To use a model to investigate, the model needs to be similar to the real world in ways that are important for what the scientist is investigating. Sometimes, what you want to model is a situation or an event. To do this, you create a model that includes the things that are part of an event and then use that to act out a situation. These are called simulations. Simulations use a model to imitate, or act out, real-life situations. Simulations imitate, or act out, what happens in real life in a way that is similar to real life but lets you examine what is happening without causing any harm or danger. Scientists use simulations when what they want to study is too big or too small, too fast or too slow, or too dangerous to investigate directly.

Airline pilots often train in aircraft simulators. They climb into a machine that looks and feels like a real cockpit. This way they can make mistakes and learn from them without harming people and property.

Materials

- plastic plates
- stopwatch
- model lava (bottle of dish soap)
- paper towels
- *Lava Flow Data* page

Design Your Procedure

As a class, spend five minutes developing a procedure for measuring lava flow. Use the materials shown on the list.

Run Your Procedure

You will have 10 to 15 minutes to run your procedure. You have everything you need to model the flow of lava.

- Use seconds to measure the time it takes for the lava to flow across the plate. Round fractions of seconds to the nearest whole second (3.51 seconds to 4 seconds).
- Run at least five trials to show that the time you measured for the lava to flow is consistent.

You will need to record your data during this investigation. Remember, recording results allows scientists to accurately report their findings. Data help others understand a scientist's work. They also help other scientists do future investigations.

Record your results on a sheet of paper. Be prepared to share your results with your class.

Communicate Your Results

Share Your Data

The last time you communicated your work, each group presented in a *Solution Briefing*. This time you will do it differently. Each group will report to the class one result (amount of time in seconds that it took for the model lava to flow across the plate). As each group reads out their results, you will chart them on a **bar graph**. You will do this by placing an "X" on the graph for every data point on your *Lava Flow Data* page.

bar graph: type of graph that uses either vertical (up and down) bars or horizontal (across) bars to show data. Data can be in words or numbers.

By creating a bar graph, you will be able to demonstrate that your class can

- accurately determine how fast lava flows, and
- reproduce the same result over and over.

Remember, your class is a team of scientists, and towns will be counting on you to save lives. Your procedure for measuring lava flow needs to be accurate in order to keep everyone safe.

Analyze Your Data

Look at the bar graph. Work as a class to answer the following questions. Discuss how your answers may help you better achieve the *Lava Flow Challenge*. Have your written procedure available as you answer the questions.

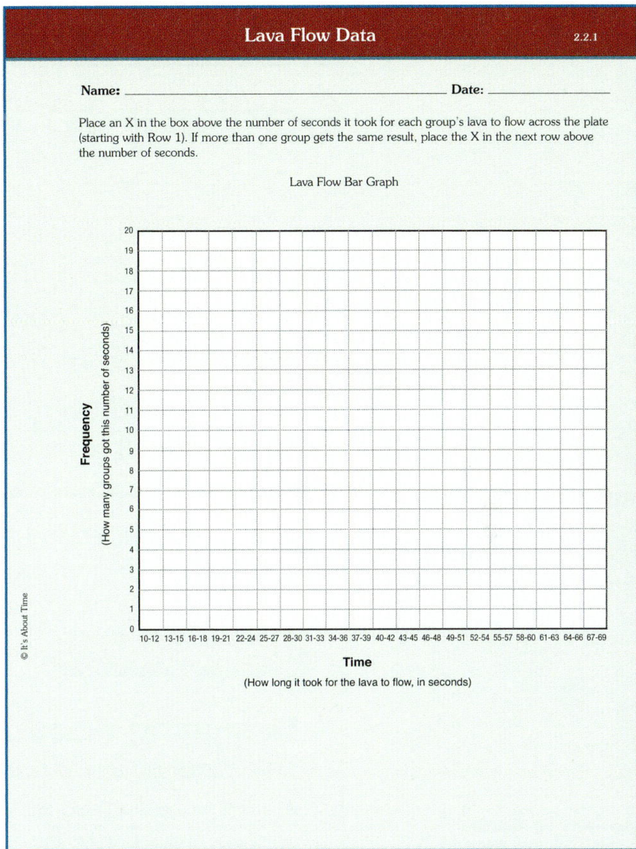
Lava Flow Data 2.2.1

Name: ______ Date: ______

Place an X in the box above the number of seconds it took for each group's lava to flow across the plate (starting with Row 1). If more than one group gets the same result, place the X in the next row above the number of seconds.

Lava Flow Bar Graph

1. Did your group have any difficulties (mistakes, spills, etc.) while running the procedures? Describe each one.
2. How similar are the results of different groups?
3. What did the distribution, or spread, of data on the bar graph look like? What do you think this says about how reliable the class's data are? Do you think the town council will trust your results?
4. Why do you think there are differences in the data from different groups?
5. What could the class do to get more consistent results?
6. Do you think the company that hired you to develop a procedure measuring lava flow for towns near active volcanoes will trust your results?

What's the Point?

Most likely, the distribution of data on your class bar graph was spread very widely. This indicates that the results are not reliable. There may be many reasons why your results varied so much. However, one of the main reasons is that different groups used different procedures.

Scientists also face this problem. To confirm the results of other scientists, they run investigations again, following the exact same steps as the original scientist. If a scientist did not provide precise procedures, results cannot be accurately duplicated.

2.3 Plan Your Investigation

Getting to a Better Procedure

Your class probably did not agree on how fast the dish soap flows. Your line plot may have shown that your class cannot produce reliable results. You will now see if you can find a way to make the results more consistent across groups.

Think about what went wrong. You were all trying to answer the same question. You all measured the flow of dish soap across a plastic plate. You all used the same unit of measurement. You all had the same materials. But every group used a slightly different procedure. You all collected data in different ways. No wonder the results were so varied.

repeatable: when someone follows the reported procedure, they get similar results.

replicate: to run a procedure again and get the same results.

soil: the loose top layer of Earth's surface, able to support life, consisting of rock and mineral particles mixed with organic matter, air, and soil.

Be a Scientist

Designing Good Procedures

Scientists only trust results that are **repeatable** by other scientists. In order for other scientists to **replicate** the results of an investigation, the procedures must be reported very precisely. Then someone else can run the procedure again and get the same results.

Making Procedures Repeatable

For example, suppose you wanted to investigate the effect of a fertilizer on the growth of plants. You would need to keep many other factors the same:

- **soil** type,
- time spent in sunlight each day,
- amount of water, and
- type of plant.

Think about one factor, water.

You would need to make sure that each group of plants got the same amount of water. They would need to be watered the same number of times. Also, they would need to be watered in the same way. You would need to follow these rules every single time you watered each plant.

trial: one time through a procedure.

precision: how close together the measured values are.

Measurement and Precision

It is also important to make the same measurement each time. You could count the number of leaves on each plant, or you could measure the height of each plant. Or, you could do both.

The tools you use can often affect measurement. You have limits to what you can see when you make a measurement. Be sure to consider how accurate the tools you use are.

Here is a checklist that you can use to make sure your measurements are consistent:

- Measure from the same point.
- Measure with the same units.
- Repeat **trials** for more **precision**.
- Start fresh. Do not compare data from before you make a change to a procedure to data collected after you make a change.
- Measure under the same conditions.

Errors and Measurement Constraints

Some groups in the class probably reported mistakes, or errors, they made as they were running their procedures. There are two kinds of errors to think about when you design procedures. Some mistakes are avoidable and others are not. For example, forgetting to start the stopwatch when the dish soap begins to flow down the plate is an avoidable mistake. These are the easiest kinds of errors to fix. You usually know when you have made this type of mistake and you can easily run the procedure again.

But some errors are unavoidable. Every measurement has error that is impossible to avoid. If you have a ruler that measures in millimeters, you sometimes cannot tell if something measures half a millimeter or a quarter-millimeter. Even if you got a better ruler, there would still be a point at which you would have to estimate. Your results can only be as accurate as your measuring tools. Because scientists have constraints on their tools, they will always make measurement errors that are unavoidable. Scientists usually keep track of possible measurement errors when they report on an investigation. This way, other scientists can judge how much to trust the results.

Revise Your Procedure

Now that you have looked at everyone's procedures and read about how to make procedures trustworthy, the class will decide on a **standardized** procedure that everyone will use to measure lava flow. A standardized procedure means that everyone will do each step the same way. That way, you will be sure that the results obtained by different people or groups can be compared.

standardized: the same.

range: the zone between the largest and smallest results.

variation: a spread of data.

First, based on the results from each group, make a list of possible sources for the differences you see. Together, as a class, design a procedure that accounts for those differences.

Work as a class to identify possible sources of error. After your class has identified the possible sources of error, revise the procedure.

Your teacher will record and display the new procedure as the class designs it. You will need to record this procedure and keep it handy.

Reflect

Review and answer the following questions. Be prepared to discuss your answers with your class.

1. What are three or four key differences between your previous procedure and the new class procedure?
2. What are you now controlling better in the new procedure?
3. What effect do you think this new procedure will have on the **range** of results across groups?

What's the Point?

The points you thought about in this section are important in the *Lava Flow Challenge.*

Every group was using a similar procedure. However, your procedures were not identical. You probably saw a wide spread of data in the bar graph. This is called **variation**. It is important to use the same procedure every time you test or measure something. Your results will then be consistent, and they will probably be repeatable. Creating a good procedure requires identifying the steps in the procedure very specifically, so it can be run the same way each time. It also requires identifying possible sources of error. This tells someone following a procedure how to do it to get repeatable results and what to be careful about as they are running it.

2.4 Investigate

Modeling Lava Flow II

Run Your New Procedure

Now that you have a new procedure, your class should be able to produce more reliable results. Your class will soon collect another set of data and produce a new bar graph. As a class, update the criteria and constraints of the challenge if you need to.

Follow your new procedure. Use the materials listed. Obtain results for 5 to 10 trials.

Record your results on the same sheet of paper you used to record your procedure. Be prepared to share your results with your class and teacher. You will have 10 to 15 minutes to perform your procedure and collect your data.

Materials

- **plastic plates**
- **stopwatch**
- **model lava (bottle of dish soap)**
- **paper towels**

Communicate Your Results

Share Your Data

Use another sheet of graph paper to make a bar graph from the new data.

As before, each group will read aloud their results. Everyone will plot them on the graph paper. Each group should report any problems they had running the procedure (e.g., mistakes, spills).

Analyze Your Data

After your class creates the second bar graph, answer the following questions together:

1. How do the results from this investigation compare to the ones from your first set of trials?
2. Did all groups get results similar to yours?
3. Do you trust these results more? Why or why not?

Revise Your Procedure

Think about and discuss how the new, more specific procedure provides a closer answer to the question: *How can we measure lava flow?*

You might find that the range of results is still too large for you to trust. If so, come up with ways to create an even better procedure.

Run this new procedure. You may need to do this part of the activity at home. As a class, plot these new results on another bar graph. Do you trust these results more? Why?

Reflect

1. What did the distribution, or spread, or data points on your latest bar graph look like? What do you think this indicates about how precise your class has been in determining a way to measure lava flows?
2. Do you think it would ever be possible, given the materials and conditions you have in the classroom, to find a procedure that would give exactly the same result each time you ran it?
3. What do you think it would take to get to an exact procedure?

Discuss your answers and how they may help you better achieve the *Lava Flow Challenge.*

What's the Point?

By developing a precise procedure for everyone in the class to use, your results became more consistent. The more consistent your class results are, the more your procedure will be trusted.

More to Learn

Lava

Lava is melted rock that reaches the surface of Earth. There is also melted rock inside Earth. Melted rock is called magma as long as it stays underground and does not reach the surface.

Scientists describe lava by its physical appearance—the way it looks. Several things affect the way lava looks:

- what it is made of (the kinds of molecules and atoms),
- the temperature at which it is flowing,
- the kinds of crystals that form, and
- the land it is flowing over.

When lava comes out of a volcanic vent, it can range in temperature from about 700°C to about 1200°C (1300°F to 2200°F). As it moves away from the vent, it starts to cool. Just like hot fudge on ice cream, lava gets more **viscous**, or thicker, as it cools. Hotter lava is generally thinner than cooler lava. This also means that lava slows down the farther away it gets from the vent where it first erupted.

The composition of lava, or what it is made of, depends on the location of the volcano. You will study more about this later on.

Basaltic Lava

Most lava is basaltic. This type of lava is thin and runny. It has lower silica content than other types of lava. Silica is the same thing sand is made of. The two types of lava commonly found in Hawaii are both basaltic.

A'a

A'a (pronounced "ah-ah") is a slightly viscous lava. This type of lava usually comes out in flows that are three to five meters (approximately 10-16 ft) thick. A'a flows are jagged and extremely sharp. A'a in Hawaiian means "hard on the feet." Even though a'a is viscous, it flows quickly.

viscous: having a sticky, thick, or gluey consistency; not free flowing.

An a'a flow appears to eat up cars and a road. The cars give you an idea of how thick the flow is.

A'a lava as it cools. Even after the lava cools, you would not want to walk on it with bare feet.

New pahoehoe cools off. The ropes are caused by the way the lava ripples just before it hardens.

The andesite in this photograph came from lava that contained more silica than basaltic lava. This flow is about 12 m (approximately 40 ft) thick.

Because dacite is more viscous, it usually does not flow far from the volcano. Dacite often forms steep-sided mounds called lava domes.

Rhyolite has beautiful flow patterns. It is made from much more viscous lava than the Hawaiian a'a or pahoehoe.

Pahoehoe

Pahoehoe (pronounced "pa-hoey-hoey") is generally less viscous than a'a. This means that it flows more easily. It usually makes flows that are less than one meter (approximately 3 ft) thick. Pahoehoe flows are very hot, over 1000°C (about 2000°F). The word pahoehoe means "like a rope." As the lava cools, it hardens into rounded swirls and rope-like patterns with a very smooth surface.

Andesite

Andesite has slightly higher silica content than basaltic lava and is more viscous, meaning it is more resistant to flow. It sometimes cools and hardens into large blocks, sometimes called block lava.

Dacite

Dacite has higher silica content than andesite and is more viscous.

Rhyolite

Rhyolite is very viscous lava with high silica content. It is found in many of the lava flows in California and the western United States.

When you modeled the viscosity of lava, you modeled only one type of lava flow. But, if you had more time, you would want to know what type of lava flow the town was likely to experience.

Basalt flows quickly because of its components, temperature, and cooling rates. Rhyolite lava has higher viscosity. It is thicker and flows more slowly.

Reflect

- How does what you just learned about different types of lava affect the way you might measure lava flow?
- Now that you know more about lava, what would you change in your measurement procedure if you had another chance to revise it?

Learning Set 2

Back to the Big Question

How do scientists work together to solve problems?

You and your classmates have been trying to develop an accurate procedure. In the end, you probably realized that it would be very difficult to develop a fully accurate procedure. But, as the different groups in the class used more similar procedures, their answers got closer to one another. You found that the way you collect data affects the answers you can find.

The first time everyone tried to determine how fast the dish soap ran across the plate, each group had different results. That is because each group used a similar, but not identical, method. The class then came up with a more standard procedure. When everyone followed this procedure, the results were closer to each other. Your data became more consistent.

There are three likely sources of inconsistent data:

- Different procedures are used for different trials.
- Factors that can affect the measured result are not carefully controlled.
- The tools used have constraints.

It is important for scientists that the results of their investigations can be trusted. They must develop very precise methods that give similar results each time. To check scientific results, other scientists repeat procedures to see if they get the same results. Scientists can trust the work of other scientists if another scientist can replicate the investigation and get the same results.

More to Learn

Rocks and Minerals

Lava is melted rock, and rock is the material that makes up Earth's solid part. Rocks are composed of **minerals**. Minerals are **inorganic** solids formed in nature, with definite chemical compositions, and unique **crystalline** arrangements of atoms. Inorganic materials do not come from living things but are formed from matter on Earth.

Minerals are either chemical elements or compounds and have specific chemical and physical properties. The properties of a mineral depend on its composition and how it was formed. Several easily observed properties commonly used to identify minerals are color, luster (how it reflects light), streak (color of the powdered mineral), hardness (how easily it is scratched), and cleavage (how it breaks apart).

A rock is a naturally formed, inorganic Earth material that holds together in a firm, solid mass. Most rocks are made of lots of grains that hold together. The grains may be mineral crystals, rock fragments, or even the solid parts of once-living things. Rocks are identified by their mineral composition and **texture**. Texture is the size, shape, and arrangement of the mineral crystals or grains in the rock.

Based on how they are formed, rocks are divided into three main categories: **igneous, sedimentary,** and **metamorphic**.

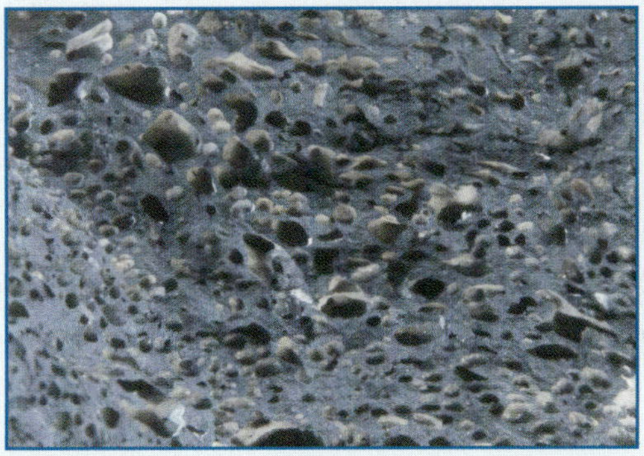

Basalts, igneous rocks formed by the rapid cooling and hardening of lava flows, are the most widespread of the igneous rocks.

Igneous rocks (from the Latin word for fire) form when hot, molten magma or lava cools and becomes solid. As the molten rock cools, mineral crystals form and increase in size until all of the material has

mineral: a naturally formed, inorganic solid, with a definite chemical composition, and a specific arrangement of atoms in a crystalline pattern.

inorganic: material that does not come from living things.

crystalline: describes the arrangement of atoms in a specific structural pattern.

texture: in this case, the size, shape, and arrangement of the grains in rocks.

igneous: one of the three main categories of rock, formed when hot, molten magma or lava cools and becomes solid.

sedimentary: one of the three main categories of rock, formed when eroded and deposited rock particles are compacted or cemented together, or from particles left behind as water.

metamorphic: one of the three main categories of rock, formed when already existing rocks are changed by heat, pressure, or chemical action.

pressure: the amount of compression force acting on a substance.

become solid. Usually, the faster an igneous rock cools and solidifies, the smaller its crystals, because they have had less time to increase in size.

Sandstone, a sedimentary rock, is mainly composed of sand-size mineral or rock grains. It forms from sediment accumulating, then compacted and cemented together.

Most sedimentary rocks form when sediments (eroded and deposited rock particles) are compacted or cemented together. Other sedimentary rocks form from particles left behind as water evaporates. Some even form from the shells of marine life. Sedimentary rocks make up less than 10 percent of Earth's crust, but they account for about three-quarters of all surface rock.

Marble, a metamorphic rock, is formed from stone that has been changed by both extremely high temperatures and tremendous pressure.

Metamorphic rocks (from the Greek words *meta*, meaning change, and *morph*, meaning form).These rocks can form from igneous, sedimentary, or even other metamorphic rocks. Great heat and **pressure** can cause the grains in the rock to change in size, shape, and density, and can cause chemical changes in the rock's minerals.

Learning Set 3

The Basketball-Court Challenge

erosion: the movement of soil or other materials from one place to another.

ingenuity: cleverness and originality.

Imagine that your school sits on top of a hill. There is no room for a basketball court. One of the parents is willing to donate some land at the bottom of the hill for the school to use as a basketball court. Everyone is excited about the idea until the school board meeting.

"Wait a minute," says the school board president. "If we build a basketball court at the bottom of the hill, how are we going to control **erosion**? What is the point of having a basketball court if it is going to be covered in dirt and mud all the time? We do not have the money to spend on a landscape report."

The school board votes not to accept the donation because of the potential erosion problem. You and your friends are very disappointed. You talk to your science teacher, and she says that erosion is a problem that can be controlled. It might take **ingenuity**, but you should be able to figure out something.

Your entire class goes to the next school board meeting. Your teacher proposes that the class investigate erosion and suggest ways it can be controlled. The school board agrees to this plan. The next school board meeting is in three weeks. If you can provide them with enough data to make an informed decision about managing erosion, they will consider accepting the land donation to build the basketball court.

Union School District
Mary Chalmers, Superintendent
27 Courthouse Square
Springfield, IL 62700

Dear Science Class:

We are delighted that your class has taken on the challenge of studying erosion control around the proposed basketball court. We will expect your report within two weeks with recommendations for erosion control on the hill above the court.

The land in question is an empty lot at the foot of a hill. There are houses on both sides of the site. You will have to make sure that none of your erosion-control measures affect these houses. We look forward to you presenting your recommendations to the school board at our next meeting.

Please use the information we have included to help you understand the requirements. Remember, you will have to be careful not to cause any damage to the properties on either side of the court.

The Proposed Basketball Court

Dimensions of court	length 28 m (about 92 ft) width 15 m (about 49 ft)
Vertical height of hill	10 m (about 33 ft)
Required distance from base of hill to court	5 m (about 16 ft)
Distance to houses on either side	12 m (about 39 ft)
Size of houses	30 x 10 m (about 98 x 33 ft)

The school board is grateful that you are able to take on this project. Good luck, and let us know if you need more information.

Sincerely,

Mary Chalmers

Mary Chalmers
School Superintendent

3.1 Understand the Challenge

Thinking About Erosion

You are going to begin by identifying the criteria and constraints of this challenge. You will then walk around your school, looking for examples of erosion. Understanding what erosion is and what causes it will be important for addressing the challenge. For now, think about erosion as movement of soil or other ground material from one place to another.

Identify Criteria and Constraints

It is always a good idea, before beginning to address a challenge, to make sure you understand the challenge. One way to do that is to identify the criteria and constraints. Remember that criteria are what you need to accomplish, and constraints are limitations on your solution. For the *Basketball-Court Challenge*, your criteria are what you need your erosion-control method to be able to do. The constraints on your solution are what you have to keep in mind and be careful about as you work on a solution. Review the letter that you received from the school superintendent, and record the criteria and constraints you identify. Then, as a class, list and discuss the criteria and constraints for this challenge.

Once you are aware of the criteria and constraints for your design challenge, you can decide which ideas are worth spending more time on and which ones are not.

The Basketball-Court Challenge

Criteria	Contraints
The solution will keep mud and dirt from sliding down the hill onto the basketball court	

The Erosion Walk

You will be taking a walk around your school and looking for examples of erosion. Look for things you might not usually notice. These might include a pile of pebbles on the side of the road or small gullies formed by a recent rainfall. You do not have to go very far to find examples of erosion.

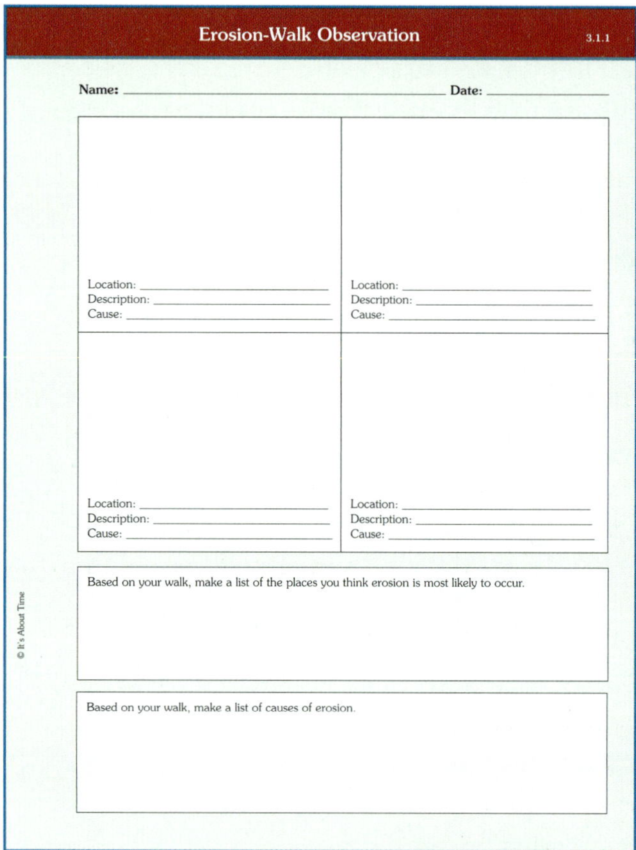

Erosion-Walk Observation 3.1.1

Name: ______ **Date:** ______

Location: ______
Description: ______
Cause: ______

Location: ______
Description: ______
Cause: ______

Location: ______
Description: ______
Cause: ______

Location: ______
Description: ______
Cause: ______

Based on your walk, make a list of the places you think erosion is most likely to occur.

Based on your walk, make a list of causes of erosion.

© It's About Time

Working with a partner, identify at least five examples of erosion. Also identify at least two examples of places where erosion should have happened but did not. Try to find examples that other students have not noticed.

It helps to notice small details. Look for clues, such as dirt that looks different than other dirt around it or rocks piled together. It is important to remember that the effects of erosion can be small or large. Erosion may cover only a few inches of ground, or you might see large areas where soil is washed away. Look carefully as you walk around.

You should also look for places where erosion should have happened but did not. For those places, try to identify what prevented the erosion. If you have any questions about how erosion happens or does not happen, be sure to record them so you will remember.

Recording Your Observations

Record your examples on *Erosion-Walk Observation* pages. You will be able to fit four observations on each page, so each pair of students will have room to record eight examples. For each example, make a sketch of the eroded area. If you can see where the eroded material came from, your sketch should also include the path the material traveled and what was formed when it was deposited. Record the location of your example, and describe it in words. Try to figure out what caused the erosion, and record that as well. You may need to look closely at the area around where you found the example of erosion to figure out how it happened. For your examples of where erosion did not happen, figure out and record what prevented the erosion. Record enough so that you will be able to share your observations with others. You and your partner may agree or disagree. If you disagree, write enough so that you remember what you disagreed about.

You will need a hard surface to write on as you complete your *Erosion-Walk Observation* page, so take a book or workbook with you to lean on. If you have a camera, bring that too.You should also have a pencil and your *Erosion-Walk Observation* page.

Conference

After you return from the erosion walk, share your observations and ideas with your group about what caused your examples of erosion. Make sure everybody has a chance to share. As a group, select two examples of erosion, and make your best guess about how each happened. For example, you may have found a small ditch carved out in a flat area. You might think that the wind slowly carried particles of dirt away from the area. Your guess may not be correct, but do your best based on what you know so far.

Then select an example of a place where erosion did not happen. Try to come to an agreement about why it did not occur.

When you are finished, try to answer the two questions at the bottom of your *Erosion-Walk Observation* page.

- Where is erosion most likely to happen?
- What causes erosion?

In the time you have left, discuss which of your answers you are sure about and which you are less sure about. Discuss what you think you still need to learn to fully understand your observations.

Create the *Project Board*

When you work over a long period of time on a project, it is important to keep track of what you have accomplished and what you still need to do. Throughout *PBIS*, you will be using a *Project Board* to do that. A *Project Board* gives you a place to organize your ideas, what you need to investigate, and what you are learning. Reading the box, *Introducing the Project Board*, will give you a better idea of what a *Project Board* is and what you will use it for.

Project Board: a space for the class to keep track of progress while working on a project.

To get started on this *Project Board*, identify the important science question you need to answer. To design erosion-control measures for a basketball court, you need to understand the answer to these questions: What causes erosion and how can it be controlled? Write these questions on your *Project Board.* Your challenge is to design an erosion-control plan to manage erosion around the proposed basketball court. Add the challenge to the top of your class *Project Board*: *How can erosion around the proposed basketball court be controlled?*

The erosion walk was meant to help you recognize what you understand about erosion. It also helped you think about what you do not understand well enough. These are the things you will record in the first two columns of the *Project Board*.

What causes erosion and how can it be controlled?
How can erosion around the proposed basketball court be controlled?

What do we think we know?	What do we need to investigate?	What are we learning?	What is our evidence?	What does it mean for the challenge or question?

Be a Scientist

Introducing the *Project Board*

When you work on a project, it is useful to keep track of your progress and what you still need to do. You will use a ***Project Board*** to do that. It gives you a place to keep track of your scientific understanding as you make your way through a Unit. It is designed to help your class organize its questions, investigations, results, and conclusions. The *Project Board* will also help you to decide what you are going to do next. During classroom discussions, you will record the class's ideas on a class *Project Board*. At the same time, you will keep track of what has been discussed on your own *Project Board* page.

The *Project Board* has space for answering five guiding questions:

- What do we think we know?
- What do we need to investigate?
- What are we learning?
- What is our evidence?
- What does it mean for the challenge or question?

Each time you use the *Project Board*, you will record as much as you can in each column. As you work through a Unit, you will return over and over again to the *Project Board*. You will add more information and revise what you have recorded. Everything you write in the columns will be based on what you know or what you have learned. In addition to text, you will sometimes want to put pictures or data on the board.

What Do We Think Know?

In the first column of the *Project Board*, you will record what you think you know. As you just experienced, some things you think you know are not true. Some things are not completely accurate. It is important to record those things anyway for two reasons:

- When you look at the board later, you will be able to see how much you have learned.
- Discussion with the class about what you think you know will help you figure out what you need to investigate.

What Do We Need to Investigate?

In this column, you will record what you need to learn more about. During your group conference, you probably came up with questions about how to explain some of your observations. You might also have figured out some things you are confused about. And you might have found that you and others in your group disagreed about some things. This second column is designed to help you keep track of things that are confusing. Record what you do not understand well yet and what you disagree about. These are the things you will need to investigate. They will be important for achieving your challenge (designing a method to control erosion).

Sometimes you are unsure about something but do not know how to word it as a question. One of the things your class will do together around the *Project Board* is to turn what you are curious about into questions you can investigate.

Later in this Unit, you will return to the *Project Board*. For now, work as a class and begin filling in the first two columns.

What's the Point?

You observed examples of erosion in your schoolyard or nearby neighborhood. Some may have been small and hard to notice and others may have been large and quite visible. They were caused by different forces in nature, such as running water, wind, and gravity.

You started a *Project Board* to help you keep track of what you understand. You also added some questions and ideas you need to investigate further. The *Project Board* is a space to help the class work together to understand and solve problems. Using it will help you have good science discussions as you work on a project.

Now that you have identified the questions you need to answer, you know what you need to do next. You need to investigate to find the answers to some of those questions.

3.2 Case Studies

What Causes Erosion?

Looking at evidence of erosion has helped you identify some of the causes of erosion. It has also helped you raise questions about what causes erosion. Other people have had to deal with erosion before. You are going to review some real-life situations in which erosion created big problems. These situations are known as **cases**. The cases you will read about have been studied by scientists and engineers. Those experts have written about these cases so others can learn how erosion happens and how to prevent it. The **case studies** you will be reading are about large-scale erosion. They will help you answer some of your questions about erosion. They will also give you some more ideas about how the erosion you identified might have happened.

Be a Scientist

Learning from Cases

Scientists and engineers often face problems that other people have faced before. Sometimes, examining those cases can help suggest ways of dealing with a new problem. A case is an example of something.

When experts work on a case that can teach others something, they often write it up as a case study. A case study includes a description of what happened, the expert's best explanation of why it happened, and lessons others can take from the experience. Cases can often help you see your problem in a new way. They can also show you how others solved similar problems. Cases can sometimes help you understand that your problem cannot be solved in the way you thought it would be. People read about cases to get ideas. Cases provide suggestions to think about and solutions to consider.

case: an example or occurrence of something.

case study: an analysis of an example or occurrence of something.

Erosion Case Study 3.2.1

Name: ______ Date: ______

Case name	
Case information: time, location	
The situation	
The setup: what was happening before the erosion? What kinds of things was the land being used for?	The problem: What erosion happened, and what problems, if any, did the erosion cause?
The erosion	
Causes: What caused the erosion?	Time: How long did it take for the erosion to happen?
Solutions	
Fixes: How did they try to fix it? Who tried to fix it?	Outcomes: What, if anything, happened as a result?
Conclusions: What can we learn from this case?	
About causes of erosion?	Anything else?

Procedure

The case studies that follow will give you some ideas about how and when erosion happens. Later, you will read more about how people have tried to reduce or limit the impact of erosion.

You identified some causes of erosion after your erosion walk. Reading these case studies will help you identify more causes of erosion.

1. Read the case study (or case studies) your teacher assigns you.
2. For each case study you are assigned, identify what happened and why. Use the prompts on your *Erosion Case Study* page to help you describe the case you are reading about well enough so others can learn from what you write.
3. Make a list of the causes of erosion that you have read about.
4. Identify an example of erosion you saw on your erosion walk that is similar to the erosion you read about. Use the case you read about to give you ideas about what caused the erosion you observed.

Case Study 1: Ocean Rock and Roll

A Case of Gradual Coastal Erosion at Work

By observing the photographs on this page, you will see one hundred years of erosion at work. In 1890, a large rock called Jump-Off Joe stood off the coast of Oregon. Look at the photographs and see how over time water wore away the land that made up Jump-Off Joe. Jump-Off Joe was made of sandstone, a very soft type of rock. The land in the background, on which the lighthouse sits, is made of basalt. Basalt is a much harder rock than sandstone.

1890

1910

1920

1970

1990

Over a very long time, wave action destroyed Jump-Off Joe. You may not think about water as being powerful. However, if you have ever been caught in a flood, tried to swim in a swift river, or been knocked down by a wave, you know how strong water can be.

Waves striking a shore carry sand particles and are very powerful. The combination of water and sand particles wears down rock faster than water alone.

Jump-Off Joe was a very large rock in 1890. Most likely, it was once even bigger. By observing the series of pictures taken over the course of 100 years, you can see that the rock has been weathered, or worn away. Erosion carries the weathered particles away and deposits them in another location.

The Cape Hatteras Lighthouse, 1985. The ocean shoreline is just 46 m (150 ft) from the lighthouse.

The Cape Hatteras Lighthouse, 2000. The lighthouse was moved back from the ocean 884 m (2900 ft) in a massive engineering project.

Case Study 2: Where's the Beach?

Moving the Cape Hatteras Lighthouse

The Cape Hatteras Lighthouse was built in 1870. At that time, people did not fully understand the forces of erosion. So building the lighthouse 457 m (1500 ft) from the water seemed reasonable. For over a hundred years, the lighthouse warned ships away from dangerous waters in a part of the ocean called the "Graveyard of the Atlantic." However, after 129 years, the lighthouse itself was in danger. Coastal erosion had worn away about 396 m (1300 ft) of beach. The lighthouse now sat within 46 m (150 ft) of the very waters it had warned so many sailors to stay away from. In 1999, the lighthouse was moved 884 m (2900 ft) back to save it from falling into the ocean.

The Cape Hatteras Lighthouse is located on the Outer Banks of North Carolina's barrier islands. Barrier islands are found all along the eastern coast of the United States, parallel to the mainland's shoreline, and along many other shorelines around the world.

Seen from above, a view of the barrier islands off the Outer Banks of North Carolina.

These long, sandy islands protect the mainland from the winds and pounding waves of the sea. But the islands are constantly changing, eroding in one place and building up in another. This is the result of waves, currents, winds, storms, and a rising sea level.

Barrier islands are often in need of protection from the forces of erosion. After all, if they were to disappear, mainland shorelines would be defenseless against the seas. Their best defense is the dunes.

Dunes are anchored in place by the deep roots of dune plants. Most beach communities work hard to protect their dunes, asking people to stay off the dunes and not pick the dune plants. Still, barrier islands are eroding at incredible rates. Because of their beauty and recreational value, barrier islands are popular places to build homes and visit. Often, people build along these beaches without considering erosion.

Beach erosion has many causes:

- building houses and hotels near the ocean;
- a rapid rise in average ocean levels;
- the gradual sinking of coastal land;
- efforts to reduce erosion that have not worked and instead have increased erosion; and
- global warming, which will speed up the rise in sea level.

But erosion is not all bad. Without erosion, there would be no beaches, dunes, barrier islands, or bays. Bays are bodies of water found between barrier islands and the mainland. They are productive nurseries for many marine organisms.

No Perfect Solutions to Erosion Problems

Places with buildings on the edges of cliffs above beaches often have serious problems with erosion. As cliffs began to erode, as they did in the city of Miami Beach, Florida, the ocean was getting closer to the buildings. People were afraid that the buildings, including many homes, would be destroyed if the cliffs collapsed.

Many cities faced with these problems try and stop the erosion of cliffs in different ways. Several of these ways have been tried in Miami Beach, and all along the Atlantic coastline. To slow down erosion, city engineers often build barriers along beaches or into the water. Structures built include seawalls—made of concrete, steel, or wire cages filled with pebbles, or groins and jetties—different types of barriers made of rocks. They are designed to keep ocean currents from carrying away sediment and sand. Often these structures shift the movement of **sediment** or sand to other parts of the beach, causing more damage in another area.

Breakwaters, long heaps of rocks dumped parallel to the shore, reduce the strength of waves before they reach the beaches. Some people object to breakwaters because they can spoil ocean views.

sediment: solid fragments of inorganic or organic materials that come from rock and are carried and deposited by wind, water, or ice.

Jetties (left) are similar to groins, but are used to keep sand away from shipping channels. This erosion-control method helps one area of the beach but hurts another area.

Breakwaters (left), are promising solutions to the problem of beach erosion. Some people object to the way breakwaters block views of the ocean.

Another way to restore beaches is to pump lots of sand onto them through a process called "beach nourishment." Just as food nourishes bodies, the sand, taken from deep in the ocean or construction projects, helps to build up the beaches. This process is expensive and, because erosion continues to remove the sand, beach nourishment must often be repeated after several years.

Seawalls (left), made of concrete, rock, steel, or wire cages filled with pebbles, are built in many places to slow down the effects of erosion.

Case Study 3: Landslides

A Case of Gravity and Water

A landslide is the sliding downhill of loose rocks and soil. Landslides occur when gravity pulls on rocks and soil. You can see from the picture how a landslide has left a huge scar that will quickly lead to further erosion. Landslides happen when the forces holding soil or rock together are smaller than the force of gravity pulling them down. For example, fires sometimes burn the trees and brush on steep mountains. The roots of the trees and brush help hold the soil in place. Once they are gone, the soil and rocks slide down the slope.

Once a landslide has occurred, the rate of erosion by water, gravity, and wind speeds up. Erosion leads to more erosion. Rain can get into rocks and cause them to become unstable. When unstable rock and soil get wet, they get heavier. The force of gravity pulls them more strongly. Land often moves more after a soaking rain.

As a child, you may have experienced the effects of a landslide yourself. If you try to build a sand castle with very dry sand, the sand tumbles downhill as it is pulled by gravity.

When you use moist sand, the sand particles stick together and the sand castle remains sturdy.

If too much water is added to the sand, the sand will become fluid-like, and it will flow downhill. The extra water strains the forces holding the sand together, and then gravity pulls the sand downward.

friction: a force that resists motion.

Laguna Beach Landslide

In 2005, the Laguna Beach landslide in California destroyed at least 11 homes. Many people were evacuated. However, this landslide could have been predicted, according to many people..

The landslide occurred because heavy rain from months earlier accumulated in the ground. This wetness reduced **friction** between the rocks the homes were built on and the underlying ground. Water accumulated in layers deep beneath the surface. If the water content of soil becomes high enough, the soil will flow like a fluid. At the Laguna Beach location, the soil deep beneath the surface became fluid-like and began to flow downhill. Drier soil and rocks from upper layers and the surface were carried along on top of the flow. This caused the landslide to push and batter homes in its path, instead of flowing around them.

Landslides happen on steep slopes. Builders in Laguna Beach should have studied the conditions before building homes in the area. But, many of the houses in this area were built before current building laws were in place. There are a lot of areas in Southern California that have conditions similar to those in Laguna Beach.

The steep slopes and rainstorms in the coastal town of Laguna Beach, California, have resulted in serious landslides, as many people anticipated. After the landslide of 2005, over a thousand people were evacuated from 500 homes and much property was destroyed or severely damaged.

Case Study 4: The Dust Bowl

Erosion Caused by Wind

The Dust Bowl occurred in the middle region of the United States, including areas of Kansas, Texas, and Oklahoma. The Dust Bowl was the name given to a 10-year period of drought that occurred in the 1930s. During this time, many people suffered great hardships, and many died.

The Dust Bowl happened because people came to the area known as the Great Plains and started plowing and farming the land. This land was not ideal for farming, but the settlers did not understand this. They did not know how to farm the plains and did not understand the effects farming could have on the land.

drought: a long period of dry weather with very little or no rain.

Before the Civil War, when settlers first passed through the Great Plains, the area between the Mississippi River and the Rocky Mountains was dry. It did not seem worth staying there, as there was no gold to be found, and the land could not be farmed. These early settlers continued on to the west coast. On old maps, they called this area "The Great American Desert."

Settlers began arriving again in the 1880s, after a period of exceptionally heavy rains. The plains were bursting with tall grass and appeared to be ideal for farming. Few people remembered how dry the plains had been just 20 years before.

People mistakenly believed that farming itself would cause more rain to fall. They also thought that building railroads and bringing in electric wires would cause more rain to fall by changing the natural electric cycles of the air. In the 1890s, there was a short drought, but soon the rains came again. It seemed like rain was normal and droughts were unusual.

In the 1930s, the drought returned, and it stayed for 10 years. The farmers had broken up the prairie soil and plowed under the native grasses. They then planted wheat. But the wheat could not survive in a drought like the grasses could. When the wheat died, its roots no longer held the soil in place. Farms turned into deserts covered with blowing sand. Huge dust storms whipped millions of tons of soil into the air. Dust storms blew soil from Kansas all the way to New York City.

Cattle were found dead in the fields with two inches of dust coating the insides of their stomachs. People coughed up clumps of dirt from the dust they had been breathing. Many of the people left the area looking for a better life somewhere else. They became known as "Okies" since many of them came from Oklahoma. A reporter, writing about one of the largest dust storms, called the area the "Dust Bowl," and the name stuck. The wind blew fertile topsoil away. Even today, this area has not completely recovered. Unfortunately, the Dust Bowl could have been avoided if the settlers had recalled the dry history of the area, had used different farming methods, and had not overplowed and overgrazed the land.

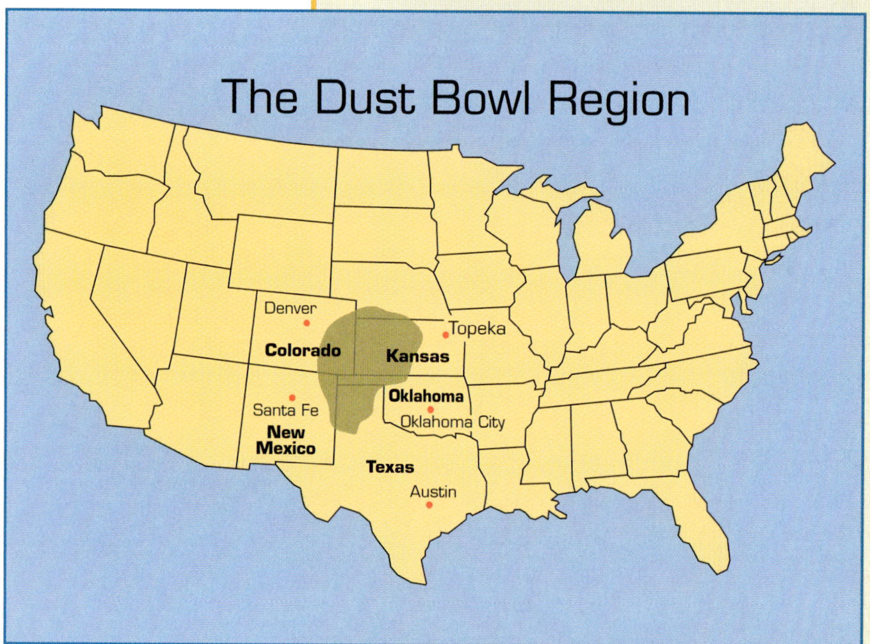

The Dust Bowl is the name given to the area of huge dust storms caused by many years of drought in the 1930s. The lack of rain, along with the plowing under of the prairie soil in the Great Plains, including Kansas, Oklahoma, and Texas, caused millions of tons of soil to blow into the air. Thousands of farms were abandoned, and many people lost their homes and suffered many illnesses from all the dust and dirt in the air. The effects of the Dust Bowl are still felt today in the Great Plains, where erosion caused by the dry winds blew away much of the fertile topsoil.

Communicate

Share Your Case Study

Each of the case studies in this section presents a different example of erosion. The environment was different in each case, as were the materials involved and the factors that caused the erosion. Since each group read only one case study, it is important that the information and lessons learned are shared with the class. Use the information you recorded on your *Erosion Case Study* page to make a presentation to your class about what you learned from the case study you read. Since your classmates will be relying on you to learn about different examples of erosion and how they occurred, it is important for you to include the following information:

- the location and time of the case;
- a description of what happened;
- how long it took for erosion to happen;
- conditions that caused the erosion to happen;
- solutions to the problem and how well they worked; and
- any negative side effects caused by the solutions.

It is important that your presentation includes answers to all of the above questions. As you listen to your classmates' presentations, it is just as important that you hear and understand the answers to the same questions. If you do not understand something, or if you think presenters left out something important, ask questions. Be careful to ask your questions respectfully and not interrupt your classmates' presentations.

Reflect

Now that you have heard about several different cases of erosion, you know a lot about erosion and what causes it. You even know some things about how to manage erosion. Taking all of the cases into account, answer the questions below. Be prepared to discuss your answers in class.

1. Based on your erosion walk and the case studies you have just read, define erosion in your own words.
2. What are two forces of nature that seem to have a significant role in erosion? What effect do they have on eroding soil, sand, and other materials?

Erosion and Weathering

Rocks exposed at Earth's surface are broken down as they interact with air, water, and living things. The breaking down of rocks into smaller pieces by natural processes is called **weathering**. Some weathering processes only cause physical changes in rock, such as changes in size or shape. Others involve chemical reactions that change the chemical composition of the rock.

Whether the process is physical or chemical, the result of weathering is solid rock broken down into pieces. Once weathering has occurred, agents of erosion remove, transport, and deposit the rock pieces.

Erosion is the process in which soil and other particles are moved from one place to another. These particles can be moved by running water, gravity, wind, waves, or even glaciers. When the particles are laid down in a new place, it is called **deposition**. Erosion can happen quickly or slowly. During a landslide, tons of soil and rock move downhill very quickly. Some erosion takes millions of years.

The Grand Canyon was formed by erosion over millions of years. Much of the erosion you see around you happens over days, months, or years.

weathering: the breaking down of rocks into smaller pieces by natural processes.

deposition: in this case, the laying down of soil or other materials in a new location.

When water moves across Earth's surface, it picks up and carries particles of soil, sand, or gravel.

Rills are small grooves cut into the soil by moving water that carries particles of soil and rock. If the erosion continues, small gullies can eventually form.

3. What are some human activities that cause erosion?

4. Describe some ways erosion can be controlled. What are some negative side effects of trying to control erosion?

Update the *Project Board*

You have learned about erosion from direct observation during your *erosion walk* and by reading several case studies. Now you will put these ideas about erosion into your *Project Board*, so everyone in the class can think about the next steps in the challenge.

As a class, review and discuss the *Project Board*. You can update or record new facts in the *What are we learning?* and *What is our evidence?* columns. When you record what you are learning in the third column, you will be answering some questions in the *What do we need to investigate?* column. You will describe what you learned from the case studies you just read.

But you cannot just write what you learned without providing the evidence for your conclusions. Evidence is necessary to answer scientific questions. You will fill in the evidence column based on the descriptions and analyses of large-scale erosion that you found in the case studies. You may use the text in this book to help you write about the science you have learned. However, make sure you put it into your own words. The class will fill in the large *Project Board*. Make sure to record the same information on your own *Project Board*.

As you read the different cases in this section, you may have thought of some new questions or ideas that you are not sure about. These can be added to the *What do we need to investigate?* column. In order to meet the challenge, you will need to learn more about erosion and the factors that affect erosion. These are some of the things you still need to investigate.

What's the Point?

When scientists and engineers are confronted with a problem, others have often had to deal with the same problem before. It is helpful to investigate and read about such similar cases. Case studies can help you understand a problem. They can also show you how others have attempted to solve similar problems and how well their solutions worked. If the solution worked, then you can try to improve upon it and make it work better. If it did not work, then you know that you have to try a different approach to solving the problem.

You had the opportunity to read four case studies. Each case showed how a different force in nature can cause erosion. You saw examples of erosion by rain, waves, gravity, and wind. Later on you will read more about some of the different methods used to control erosion. Some have been successful while others have not. This information will be very valuable to you as you work to achieve success at the *Basketball-Court Challenge*.

Beach erosion, like other types of erosion, results from a combination of factors: construction near shorelines, a rapid rise in ocean levels due to global warming, and the gradual sinking of coastal lands.

Gravity pulls soil, rocks, and other particles down slopes or hillsides, causing erosion.

3.3 Investigate

Investigating Factors that Affect Erosion

On your erosion walk and while you were reading the cases, you may have noticed that the type of soil or other Earth materials can make a difference in how and when erosion occurs. In this section, you will investigate several different types of soil and materials to see how water and gravity affect their erosion.

Your class will complete two investigations. One half of your class will investigate the relationship between particle size and erosion. The other half will investigate the relationship between steepness of slope and erosion. Each group will collaborate to interpret their observations and then share their findings with the class. In this way, you will be able to learn from one another.

Be a Scientist

Variables and Designing Experiments

When you investigate a **phenomenon**, you want to learn about the factors that influence it. In science, these factors are called **variables**. The point of most experiments is to understand how a variable affects the phenomenon you are investigating. The phenomenon you are studying is erosion. There are many variables that affect erosion; you will focus on how the type of material and slope of the land affect erosion.

phenomenon: an event or detail that can be observed.

variable: a single factor that is tested in an experiment.

Investigation 1: What Is the Relationship Between Particle Size and Erosion?

You saw in the case studies you read that water is a very powerful agent of erosion. Many of the examples of erosion you saw on your erosion walk were probably caused by water. As water runs over Earth's surface, it picks up and carries away particles of soil and other materials. Some particles are more easily carried by water than others. In this investigation, you will compare how different-sized particles are eroded by water. Particle sizes most easily carried by water will be most easily eroded.

Predict

You will be working with a mixture of sand, gravel, slate, and potting soil or native soil. Native soil is the soil you can find where you live. Which material do you think will be most easily carried by water? Which materials do you think will be more difficult for water to carry? Which material do you think will be most difficult for water to carry? For each, record your answer and why you think that.

Materials

- **1 cup pre-wetted, mixed earth materials**
- **stream table**
- **small cup with water**
- **ruler**
- ***Particle Size and Erosion* page**

Procedure

1. Obtain 1 cup of pre-wetted, mixed earth materials.
2. Dump the wet mixture near the center of one end of the stream table tray. Form it into a hill against the edge of the tray. Raise the end of the tray beneath the hill with a book.
3. Fill a clean plastic cup with water and slowly pour the water over the top of the hill.
4. Observe the erosion on the hill.
   - First, notice the sizes of the particles. Rank the materials according to the size of their particles, from largest (1) to smallest (4). Enter these numbers in the first column of your *Particle Size and Erosion* page.
   - Measure how far the particles of different materials spread beyond the hill. Record measurements in the second column of the table.
   - Observe the patterns of eroded material, and describe them in the next column.
   - Observe and describe how the hill was affected by the running water. Add this to the bottom of the page.

Particle Size and Erosion 3.3.1

Name: ______________________ Date: __________

Particle in mixture	Particle size rank (largest = 1, to smallest = 4)	Distance material spread	Patterns of eroded material	Effects on hill
Gravel				
Sand				
Slate				
Soil				

5. Drain the water from the soil, form the soil back into a hill against the edge of the tray, and repeat Steps 3 and 4. Do a total of 3 trials. Record the results for each trial.

Slope and Erosion 3.3.2

Name: ______________________ Date: ______________

Slope of tray	Distance material spread	Patterns of eroded material	Effects on hill	Other observations
No slope				
Gentle slope				
Steep slope				

Investigation 2: What Is the Relationship Between the Slope of the Land and Erosion?

The Laguna Beach case showed that water and gravity can work together to cause erosion. In Laguna Beach, there was so much rain that a cliff's soil became thoroughly soaked with water. The water made the cliff so heavy and loosened the bonds between the particles of soil so much that the force of gravity caused whole parts of the cliff to come loose and fall downhill. This is called a landslide.

Water and gravity also work together in smaller ways to cause erosion. On your erosion walk, you may have seen examples of soil or rocks that looked like they had rolled down a hill. In this investigation, you will experience how water and gravity together affect the way earth materials move. You will construct hills on different slopes and identify how each responds to water falling on it.

Materials

- **2 cups pre-wetted, fine sand**
- **filled spray bottle**
- **1 stream table tray**
- **small cup with water**
- **ruler**
- ***Slope and Erosion* page**

Predict

You will be working with hills of sand at different slopes to see how each would be affected by rainfall. On which slope do you think materials will travel the farthest when water falls on it? On which slope do you think materials will travel the least far? Why?

Procedure

1. Obtain two cups of pre-wetted, fine sand.
2. Dump the wet sand near the center of one end of the stream table tray. Form it into a hill against the edge of the tray. Leave the tray lying flat on your desk (no additional slope).

3. Spray water gently on the top of the hill about 50 times. Try to spray the water on the hill the same way that rain would fall. During spraying, notice and record any movements of the sand particles on or around the hill in the *Patterns of eroded material* column of your *Slope and Erosion* page.

4. Continue spraying until the spreading of the hill slows or stops (about another 50 sprays).

5. Observe the erosion of the hill. Observe how far the particles of sand spread beyond the hill. Observe the patterns of eroded material. Observe how the hill was affected by the running water. Record your observations on your *Slope and Erosion* page.

6. Drain the water, form the soil back into a hill against the edge of the tray, and repeat the procedure, raising the end of the tray beneath the hill about 5 cm (gentle slope) in Step 2.

7. Drain the water, form the soil back into a hill against the edge of the tray, and repeat the procedure again. This time, raise the end of the tray beneath the hill about 10 cm (steep slope) in Step 2.

interpret: to find the meaning of something.

trend: a pattern or a tendency.

claim: in this case, a statement about what a trend means.

Analyze Your Results

Finding Trends and Making Claims

Your class has now collected data about how different materials are affected by erosion and how slope affects erosion. It is now time to **interpret** those results. To interpret means to figure out what something means. Interpreting results of an experiment means identifying what happens as a result of changing a variable. What happened to each of the different types of materials when the mixture was eroded by running water? How did the movement of particles of different materials compare? What spreading patterns could be observed? How did changing the steepness of the hill affect the rate of erosion? How did the steepness of the hill affect how fast water flowed downhill? How did slope affect the distance the hill spread and the amount of sand carried to the bottom of the tray?

You will do two things to interpret your results. First, you will identify **trends** in your results. Then you will state a **claim** based on those trends. A trend is a pattern that you can see over several trails. A claim is your statement about what those trends mean. For example, you varied the slope to see if it affected the rate of erosion. Your data may have shown that the finer particles in the mixture eroded faster than the larger particles over all the trials. This is a trend. Your claim would be the statement: "Finer particles erode faster than larger particles."

Every time a scientist makes a claim, other scientists look for the evidence the scientist has for that claim. One kind of evidence is data collected in an experiment and the trends in that data. You will spend a lot of time in PBIS Units making claims and supporting them with evidence. You will be learning more about that later. For now, make sure that the data you collected matches your claim.

Make sure to record the trends you have identified on your *Particle Size and Erosion* and *Slope and Erosion* pages. Also include any claims you think you can make so you can share them with your classmates.

Communicate Your Results

Investigation Expo

You will now share with the class what you have found in an ***Investigation Expo***. Remember, no groups in the class did both investigations. Therefore, others will need your results to complete the challenge. Read the box introducing *Investigation Expos* before moving on to make yourself familiar with what you will be doing in this activity.

Be a Scientist

Introducing an *Investigation Expo*

An *Investigation Expo* is like other presentations you have done, but specially designed to help you present results of an investigation. You will include your procedure, results, and interpretations of results.

Scientists present results of investigations to other scientists which lets the other scientists ask questions and build on what was learned. Scientists may present results by making posters and setting them up in large rooms at meetings with other scientists and their posters. They also give presentations about their investigations and results in front of large audiences of other scientists. Their presentations usually include visuals (pictures), showing all the important parts of their procedures and results.

To prepare for an Investigation Expo, you will usually make a poster that includes the same items that scientists' do.

Investigation Expo: **presentation of the procedure, results, and interpretations of results of an investigation.**

fair test: things being compared are tested under the same conditions, and the test matches the question being asked. All variables, aside from the variable you are investigating, are kept the same.

These include:

- questions you were trying to answer in your investigation;
- your prediction;
- your procedure and what makes it a **fair test**;
- your results and how confident you are about them;
- your interpretation of the results (conclusions).

If you think the test you ran wasn't as fair as you had planned, report on how you would change your procedure if you had a chance to run the investigation again.

As you look at the posters and listen to other groups present their work, start with the groups that did the same investigation you did. Notice the similarities and differences in what they found and in their conclusions. If another group got different results, try to decide whose results are more accurate, yours or the others. If another group had different conclusions from yours, decide whether or not you agree with their conclusions and why.

When you look at the posters and hear the presentations of the groups that did the other investigation, make sure you get answers to all of these questions:

- What was the group trying to find out?
- What variables did they control as they did their procedure?
- Is their data consistent?
- Did they run their procedure the same way every time?
- What did they learn?
- What conclusions to their results suggest?
- Do you trust their results? Why or why not?

During the presentations, make sure you understand the procedure each group followed and that you agree with each group's conclusions. If you do not hear answers to all the questions, if the answers are not clear, or if you think a group made a mistake, ask questions. Be sure to ask your questions respectfully.

Be a Scientist

Different Kinds of Variables

As you designed and ran your experiment, there were several kinds of variables you worked with:

- One that you changed or varied in your experiment. This is called the **independent variable** (or **manipulated variable**).
- Some were ones you worked hard to keep the same (constant) during every trial. These are called **control variables**.
- Some were ones you measured in response to changing the manipulated variable. These are called **dependent variables** (or **responding variables**). Their value is dependent on the value of the independent or manipulated variable.

Experiments are a very important part of science. When scientists design experiments, they think about the things that might have an effect on what could happen. They then identify exactly what they want to find out more about. They choose one factor as their independent (manipulated) variable. This is what they change to see what happens. They have to keep everything else in the procedure the same. The variables they keep the same, or hold constant, are control variables. Finally, there are factors they measure. These are the dependent (responding) variables. If they have designed a fair test, they can assume that changes in the dependent (responding) variables result from changes made to the independent (manipulated) variable.

If you ran Investigation 1, your independent (manipulated) variable was the type of material. If you ran Investigation 2, your independent (manipulated) variable was slope. In both experiments, your dependent (responding) variables were the distance the materials spread, the patterns of erosion, and the effects on the hill. Everything else, including the shape and size of the test container, amount of material tested, amount of water poured on each sample, and the way the water was poured, were control variables. To be sure that what was measured (the dependent or responding variable) was dependent on what was changed (the independent or manipulated variable), it was important to keep the controlled variables exactly the same every time a trial was run.

independent (manipulated) variable: in an experiment, the variable the scientist intentionally changes.

control variables: in an experiment, the variables kept constant (not changed).

dependent (responding) variables: in an experiment, the variables whose values are measured; scientists measure how these variables respond to changes they make in a manipulated variable.

Reflect

Answer the following questions. This will prepare you for a class discussion about what you now know that will help you achieve the *Basketball-Court Challenge*. Be prepared to discuss your answers with your class.

1. What variable were you investigating in your experiment? What were you investigating about that variable? How did you vary it to determine its effects?
2. List all of the variables you tried to hold constant in your experiment.
3. How many trials did you perform? Why did you perform that number of trials? Was this a good number of trials?

4. For those who did *Investigation 1*: How consistent was your set of data? Why is consistency in repeated trials important in an experiment?

   For those who did *Investigation 2*: How consistent was your data with the data of other groups who ran the same investigation? Why is it important for your data to be consistent with the data collected by other groups?
5. How useful was your data in determining the affect of your variable on erosion?
6. What do you think you now know about the effects of particle size on erosion that will help you design a way to control erosion at the basketball court? What do you think you know about the effects of slope on erosion that will help you control erosion at the basketball court?

Slope, Particle Size, and Erosion

Erosion moves soil and other particles. Force is needed to move anything; the main driving force of erosion is gravity. Gravity can move sediments by acting on them directly. Pieces of rock on cliffs and steep slopes, broken loose by weathering, fall or slide downhill under the direct influence of gravity. Gravity can also move sediments by acting on them through agents of erosion. If water runs downhill under the direct influence of gravity, the running water can then exert an indirect force on rock particles in its path, causing them to move. The running water is an agent of erosion. Other agents of erosion include winds, glaciers, waves, and ocean currents.

The faster water moves, the more force it can exert. With more force, water can move more rock particles and larger ones. The speed at which water flows downhill is directly affected by the slope, or steepness, of the land. The steeper the slope, the faster the water flows downhill, and the greater its power of erosion.

Water moving at different speeds can move different-sized particles. If water moves at 50 cm/s, it exerts enough force to move sand particles (and anything smaller) but not pebbles. At that speed, with water flowing over a mixture of sand and pebbles, the sand will be carried downstream, but the pebbles will be left behind. This way, running water can cause a mixture of different-sized particles to become sorted, or separated, according to their size.

What's the Point?

Your class completed two investigations to answer two different questions. One half of the class collected data about how particle size affects erosion of soil and other materials. The other half of the class collected data about the effects of slope on the erosion of soil and other materials. Each group then interpreted their results by identifying trends in the data and stating a claim based on those trends. When everyone was finished, each group shared what they found in an *Investigation Expo*. By sharing results, everyone was able to get the information needed to answer both questions. This is the way scientists work. Presenting results of investigations to other scientists is one of the most important things they do. This lets other scientists build on what they learned. You interpreted the data from your investigation. The trends you found and the claim you made will help you in achieving the *Basketball-Court Challenge*.

3.4 Explain

Create an Explanation

After scientists get results from an investigation, they try to make a claim. They base their claim on what their evidence shows. They also use what they already know to make their claim. They explain why their claim is valid. The purpose of a science explanation is to help others understand:

- what was learned from a set of investigations or case studies, and
- why the scientist reached this conclusion.

Later, other scientists use these explanations to help them explain other phenomena. Explanations can also help them predict what will happen in other situations.

You will do the same thing now. Your claims and explanations will be about what causes erosion. Each group will use the examples found on your *Erosion Walk*, information from the case studies you have read, and evidence from the experiments you just completed to make a claim about why erosion happens. You will then create an explanation to support your claim. You will be reporting your explanation to your classmates. With a good explanation that matches your claim, you can convince them that your claim is valid.

Because your understanding of the science of erosion is not complete, you may not be able to fully explain the causes of the erosion you observed. But you will use the evidence you have collected and what you have read to come up with your best explanation. Scientists finding out about new things do the same thing. When they only partly understand something, it is impossible for them to form a "perfect" explanation. They do the best they can based on what they understand. As they learn more, they make their explanations better. This is what you will do now and what you will be doing throughout PBIS. You will explain your results the best you can based on what you know. Then, after you learn more, you will make your explanations better.

Be a Scientist

What Do Explanations Look Like?

Making claims and providing explanations are important parts of what scientists do. An explanation connects three parts:

- Claim—a statement of what you understand or a conclusion that you have reached from an investigation or set of investigations.
- Evidence—data collected during investigations and trends in that data.
- Science knowledge—knowledge about how things work. You may have learned this through reading, talking to an expert, discussion, or other experiences. Science knowledge comes from investigations others have done over many years. It is knowledge that scientists agree about.

An explanation is a statement that connects the claim to the evidence and science knowledge in a logical way. A good explanation tells what causes the statement in a claim. The best scientific explanations use agreed-upon science knowledge in a logical way to support a claim. These kinds of explanations can usually convince others that a claim is valid.

For example, suppose you live in a city in the USA that gets cold and has snow in the winter. It is fall. You see a lot of birds flying past your home. You wonder why so many birds are flying by. You have learned that many birds cannot live in cold places. They fly to warm places (usually south) to spend the winter. You wonder if these birds are flying by your home on their way to a warmer place. You take out your compass and observe that the direction they are flying is south. You conclude that the birds are flying past your home to a warmer place where they will spend the winter. Look at how you can form an explanation.

Your claim: The birds flying past my house are flying south for the winter.

Your evidence: The birds are flying in a southern direction. (You have observed and measured that using a compass.) It is autumn.

Your science knowledge: Many birds cannot live in cold weather. Birds that cannot live in cold weather fly to warmer climates when the weather begins to cool. They stay there for the winter.

Your explanation: The birds flying past my house are flying south for the winter. Many birds cannot live in cold weather. Birds that cannot live in cold weather fly to warmer climates when the weather begins to cool. These birds are probably birds that need to go to warmer places in the winter. They are flying south to find a warmer place to stay while it is winter here.

An explanation is what makes a claim different from an opinion. When you create an explanation, you use evidence and science knowledge to back up your claim. Then people know your claim is not simply something you think. It is something you have spent time investigating. You have found out things that show your claim is likely to be correct.

Explain

Writing an Explanation

Here is an example from the *Boat Challenge* that might help you understand more about writing an explanation. When you worked on that challenge, you saw that boats designed to push down on a large area of water and to hold air in their structures were able to keep afloat even with the weight of the keys. You read that a flat piece of foil has more buoyant force than a foil ball because a greater area of water is pushing up on it. You also read that having air as part of the structure of the boat decreases the overall density of the boat and its cargo and increases the buoyant force of the boat. Look at the claim and the explanation that could have been created by a group of students who worked on the *Boat Challenge*.

Some boats, such as this barge, are designed to carry huge amounts of cargo and still float easily through the water. If the boat is less dense than the water it displaces, the boat will float.

Look at the claim. Remember that a claim is a statement of proposed fact that comes from an investigation.

Claim – a statement of what you understand or a conclusion that you have reached from an investigation or a set of investigations.

A boat with a larger surface area and a hollow compartment to hold air will float better than a boat with a smaller surface area and no compartment to hold air.

This claim comes from looking at the trend they found in the data they collected.

Evidence – data collected during investigations and trends in those data.

Flat foil boats with hollow compartments were able to hold 8 keys and to float for 20 seconds. Boats that were not flat or did not have hollow compartments did not hold as many keys and sometimes sank.

They then read about the science of buoyant force and summarized the science knowledge they gained.

Science knowledge – knowledge about how things work. You may have learned this through reading, talking to an expert, discussion, or other experiences.

As the surface area of a boat increases (when its weight stays the same), it pushes down on a larger area of water. When this happens, the boat is in contact with more water molecules that push up on the boat, and the water can better support the boat. If the force pushing up from the water is greater than or equal to the force pulling down on the boat by Earth, the boat will stay afloat.

When you place a boat in water, it pushes away some water. When the overall density of the boat and its cargo are less dense than the water that displaces it, the boat will float.

Air has a much lower density (mass per volume) than water. If air is included in the overall materials making up the boat, that air could decrease the overall density of the boat and help make it more likely to float.

Here is a logical statement that ties the claim, evidence, and science knowledge together. Notice how the claim, the evidence, and the science knowledge are all part of this explanation.

> ***To hold a lot of mass, we recommend designing a boat that covers as much surface area as possible and that has places in its structure that hold air. Our data showed that the boats with larger surface areas and the ability to hold air remained afloat even with the increased weight of the cargo. We know that, to increase the buoyant force of an object, you can spread the mass of the object over a greater area. The larger surface area is in contact with a greater number of water molecules that exert an upward push on the boat, providing more support. We also know that air has a very low density and can decrease the overall density of an object. You can decrease the downward push of the boat by decreasing its density. As long as the push down from the boat is not greater than the push up from the water, the boat will float, even with cargo.***

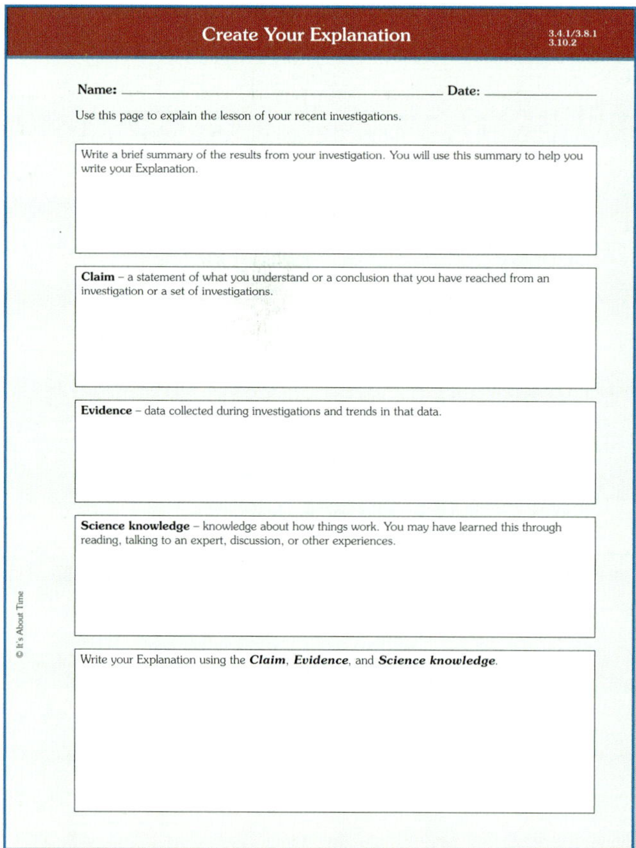

Create Your Explanation 3.4.1/3.8.1 3.10.2

Name: ______________________ Date: __________

Use this page to explain the lesson of your recent investigations.

Write a brief summary of the results from your investigation. You will use this summary to help you write your Explanation.

Claim – a statement of what you understand or a conclusion that you have reached from an investigation or a set of investigations.

Evidence – data collected during investigations and trends in that data.

Science knowledge – knowledge about how things work. You may have learned this through reading, talking to an expert, discussion, or other experiences.

Write your Explanation using the ***Claim***, ***Evidence***, and ***Science knowledge***.

© It's About Time

This may seem like a long explanation. Long explanations are not always needed. However, seeing this explanation may help you as you try to write an explanation. You will use a *Create Your Explanation* page, similar to the one shown, to help you with explanations. It will give you space to write your claim, your evidence, and your science knowledge. It will also remind you what each of these is.

Communicate

Share Your Explanation

When everyone is finished, you will share your explanations with the class. As each group shares theirs, record the explanation. You might also create a poster for the classroom that has the full set of explanations on it. You will have an opportunity to revise your explanations after you learn more about what causes erosion and how it can be managed.

What's the Point?

Science is about understanding the world around you. Scientists gain understanding by investigating and explaining. The results of investigations are useful in making sense of and organizing the world. To help others better understand what they have learned through their investigations, scientists must communicate their results and understandings effectively. Scientists make claims about the phenomena they investigate. They support their claims with evidence they gather during investigations. They also read science that others have written about. They combine all of that together to create explanations of their claims—statements about why their claims are so. Other scientists carefully examine these explanations. They discuss them with each other. They try to decide if the explanation is complete enough for them to be sure about whether the claim is valid. Scientists accept a claim as valid when many different scientists agree. The evidence and their science knowledge must justify the claim. Scientists also help each other make their claims and explanations better.

Throughout this school year, you will investigate a variety of phenomena. You will apply what you learn to solving *Big Challenges* and answering *Big Questions*. You will be asked to create explanations. Every explanation you write will include a claim, evidence, and science knowledge. As you move through each Unit and learn more, you will create new explanations. You will have the opportunity to edit and improve the explanations you created earlier. Just as you iteratively improved your boats, you will iteratively improve your explanations.

You will also use explanations you create to help you predict what will happen in new situations. For example, now you know that rain, waves, gravity, and wind are all causes of erosion. That means you can probably predict what might cause erosion around the proposed basketball court. Make sure you can make that prediction. Making that prediction successfully will help you know that you understand the science you have been learning.

3.5 Case Studies

What Are Some Ways Erosion Can Be Managed?

You are probably aware that the problem surrounding your proposed basketball court has to do with eroding soil material due to water and gravity. You are now ready to start looking for ways to control the erosion. Knowing about erosion-control methods that have been used by others may help you solve your problem at the basketball court.

You will read about different erosion-control methods and see what you can find out. You may also have some ideas about how to control erosion based on what you have seen in your own neighborhood or around your community. Real-life examples that you are familiar with can be cases, too. You will have the opportunity to share these examples after everyone has read the cases that are coming up, and your class is ready to update the *Project Board*.

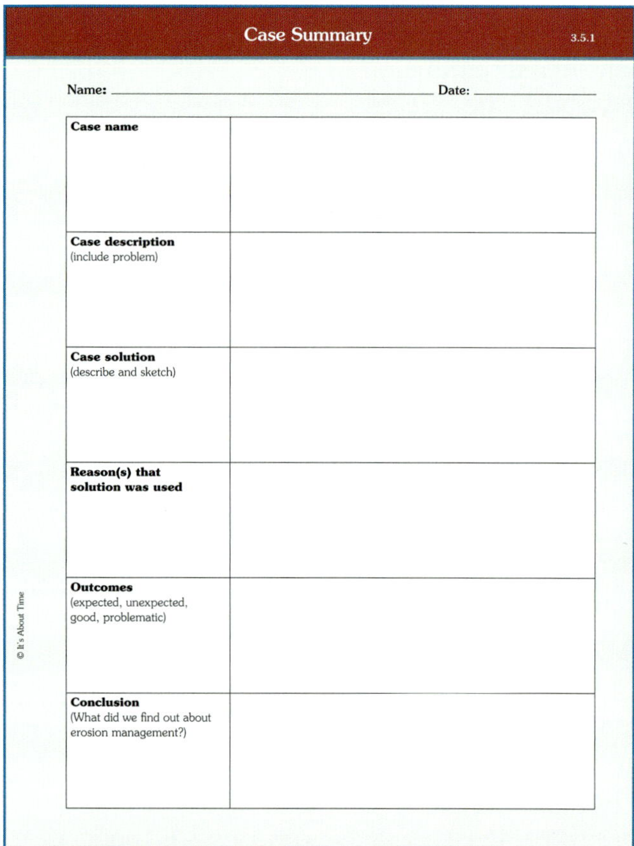

Case Summary 3.5.1

Name: ____________ Date: ____________

Case name	
Case description (include problem)	
Case solution (describe and sketch)	
Reason(s) that solution was used	
Outcomes (expected, unexpected, good, problematic)	
Conclusion (What did we find out about erosion management?)	

Coastal erosion is a natural process that results from precipitation, wind, and the constant movement of water, sand, and rock. Because communities have designed buildings very close to beaches and on cliffs above oceans, coastal erosion has damaged many structures and has put many buildings in danger of collapse.

On the next few pages, you will find pictures and text about erosion control. You will soon have the chance to model some of these techniques. Read all of the case studies, and work with your group to complete a *Case Summary* page as you review each case. Some of these cases deal with problems that are similar to the one you are facing at the basketball court. You should look for these similarities and pay attention to how the problem was solved.

Case Study 1: Boggy Creek Tributary

The storm discharge from Boggy Creek Tributary at the Poguito Street culvert (a drain passing under a road) had washed away a portion of the land on either side of the channel. The land was quickly eroding away, and the channel was getting very close to a nearby house. This project rebuilt and strengthened the bank, or side, of the channel, using limestone blocks to build a wall. The wall prevented the water from further eroding the soil. The yard was restored and the home is now protected.

You can see the damage due to erosion before the problem at the Poguito Street culvert was corrected.

Building a wall made of limestone blocks restored the area and prevented any further erosion.

terracing: cutting a series of raised steps into a slope.

Case Study 2: Tannehill Branch Creek

Water and gravity were eroding the banks of the Tannehill Branch Creek at Lovell Drive. The town needed to make 122 m (about 400 ft) of the bank more stable. They needed a way to prevent more soil from sliding downhill into the stream and being carried away by the water. They added structures that controlled the slope, or steepness, of the channel. The bank of the channel was rebuilt using limestone boulders and bundles of compacted soil and brush, called wrapped soil lifts. Both the boulders and wrapped soil lifts were used for **terracing** to build up the bank of the creek. Terracing, or building a series of steps into a slope, has been used for many years as a method of erosion control. Native grasses were also planted on the bank.

Water and gravity were wearing away the soil along the banks of the Tannehill Branch Creek.

Limestone boulders and wrapped soil lifts were terraced to build up the bank of the creek.

The roots of native grasses planted along the creek banks hold the soil in place.

Case Study 3: Shoal Creek

There were several problems along the Shoal Creek bank. High waters from storm flows, obstructions in the stream channel, and erosion along the hillside banks contributed to the erosion problem near Pembrook Drive. A large oak tree had slid down into the creek and rested on the bottom of the channel. Another live oak, as well as a house, was threatened by further erosion. In this situation, the creek bank was rebuilt with concrete, wrapped soil lifts, and native grasses. After the project was completed, the bank was stable, and the natural stream setting was attractive. A house and oak tree were also protected.

Shoal Creek before the channel was cleared and the banks made stable.

Erosion-control methods used at Shoal Creek made the stream banks stable, preventing further land loss.

Case Study 4: Little Walnut Creek

In the early 1980s, a property owner constructed a stone wall to hold back the soil and prevent it from sliding downhill into Little Walnut Creek. In December 2000, this retaining wall collapsed into the creek. This put two homes located near the stream and about 5 m (about 18 ft) above the creek bed in danger. The property next door also had a stone wall that was a concern. City leaders provided the money to rebuild the wall and protect the homes. About 107 m [350 ft] of stream bank was rebuilt with limestone boulders, plantings of native grasses and trees, and special soil reinforced with synthetic materials for strength. The completed project protected the three homes and made a beautiful, natural stream setting.

The collapse of the original stone wall made the stream bank vulnerable to erosion by water and gravity.

The completed project along Little Walnut Creek successfully reinforced the stream banks and prevented further erosion.

Case Study 5: Fort Branch Watershed

On the Fort Branch Watershed at Woodmoor Drive, there is a confined drainage channel. The sides of the channel were cut down, and the banks were eroded. The town reconstructed the bank and created a winding, more natural-looking channel using logs made of compressed natural materials, such as shredded coconut husks, and wrapped with rope webbing. The slope of the channel was controlled using rocks.

You can see the damage from erosion at the Fort Branch Watershed before the channel was rebuilt.

Effective erosion-control methods created a natural-looking channel and prevented further erosion.

Stop and Think

For each case study you read, answer the following questions:

- What erosion problem were they addressing?
- What erosion-control methods were used to address the problem?
- Why did they think the chosen method would be a good one to use?
- What happened?
- In what ways did the erosion control work as planned, and in what ways, if any, did it create new problems?

Discuss the answers in your group, and be prepared to share your answers with your classmates.

Reflect

Working with your group, identify ways of using some of the erosion-control methods you have just read about. Some of the *Reflect* questions will have you thinking about where you have seen these methods used. Others will help you think about what might be useful around the basketball court. Be prepared to discuss your answers with your class.

1. Which of the erosion-control methods you read about have you seen used in your neighborhood or community? How was each used? Why do you think each was used in that place? Draw a diagram to help you communicate how and why each was used.
2. For each of the erosion-control methods you have identified, which of them might be useful at the basketball court? Describe why and how it might be used. What do you think will result from using it? Think about good things and problems that might result. Draw a diagram to help you communicate.
3. Choose another erosion-control method you think might be useful around the basketball court. Describe how that erosion-control method is used in one of the case studies. Why do you think you could use this approach for the basketball court? How might you use it? What do you think will result from using it? Think about improvements and problems that might result. Draw a diagram to help you communicate how and why it might be used.

4. Are there any other erosion-control methods used in the case studies that might be useful for the basketball court? If so, answer the same questions about them.

Update the *Project Board*

Earlier, you began a *Project Board* centered on the idea of learning about what erosion is and how to manage it. Now you have read some case studies about how others have solved their erosion problems. You know more about the factors that cause erosion and different ways to stop it. You are now ready to fill in the *Project Board* more completely.

What is most important to add to the *Project Board* right now are your ideas about how you might control erosion in the *What do we think we know?* column. It is also important to add what you still need to find out about erosion control to address the challenge in the *What do we need to investigate?* column. Identify erosion-control methods you have read about. Then identify what else you need to know about each of those methods to design an erosion-control method for the basketball court.

The *Project Board* is a great place to start discussions. You may find that you disagree with other classmates about what you know about an erosion-control method. If so, put a question about it in the *What do we need to investigate?* column. Discussing disagreements is a part of what scientists do. Such discussions help scientists identify what they or others still do not understand well and what else they still need to investigate to understand more fully.

What's the Point?

As you read each case study, you found some similarities between these situations and the basketball-court situation. You were able to see how others solved erosion problems caused by water and gravity. Retaining walls, terracing, and drainage systems can be used to direct the flow of water and keep water flowing in places where it cannot cause damage. Planting native grasses and trees has also been used to control erosion. Plant roots anchor the soil and other materials, preventing them from being carried away. In the cases you read, erosion-control methods were combined to solve erosion problems.

3.6 Plan

Model Erosion Control

You have just identified ways you think erosion can be controlled. It is now time to investigate how well each of these methods might work to control erosion around the basketball court. Each group will investigate a different erosion-control method by building a model of it and then running a simulation using your model. To do this, you will need to build a model of the basketball court site and the erosion-control method assigned to your group. Each group will receive a small container of soil to use in building their models. The soil in the container is similar to that surrounding the basketball court.

Architects usually build scaled-down models before the actual building begins. This way, they can get feedback on their designs and identify any potential problems.

Your first step will be to design your model so you can investigate the application of your erosion-control method to the basketball court site. Then you will plan your simulations. After discussing these with the class, you will build and test your erosion-control method.

Remember that your goal right now will be to learn how someone might use your assigned erosion-control method at the basketball court site. You will need to identify where to place it, how to place it, and what is likely to happen as a result of using it. You should aim to identify both its strengths and weaknesses when used around the basketball court. In a later section, you will have a chance to design what you think will be the best approach to managing erosion around the basketball court. That solution might combine several of the methods you are investigating in this section and the next.

Be a Scientist

Investigating with a Model

In the *Lava Flow Challenge*, you read about and discussed using models to help you investigate problems. Your class reviewed how a model makes it possible to test ideas and get scientific results when it is not very easy or even possible to study the real-life situation. For example, it is very difficult for scientists to complete investigations about galaxies millions of light-years away. They cannot travel there or manipulate objects in the galaxy. So scientists create models that are similar to the actual galaxy, and they investigate using those models.

Remember that models are representations of something in the real world. Simulations use models to imitate, or act out, real-life situations. You are going to investigate erosion-control methods by building a model of the basketball court and your erosion-control method. Then you will use your model to simulate the erosion-control method you are investigating. It would be too expensive to try out all the erosion-control methods at the real basketball court site. Your model and simulation will help you determine how effectively your erosion-control method will be at controlling erosion at the basketball court site and how to make it effectively do its job.

In the next few sections, each group will build a model that includes the basketball court, the hill above it, and one of the erosion-control methods. Using their model, each group will then simulate erosion control around the basketball court.

Remember that a simulation can only teach you something if it is run on a model that accurately represents the conditions of the real-life environment. Models can never match the real world exactly. For one thing, models are usually scaled down. "Scaled down" means smaller than the actual size. Models also do not have all the details of the real world. For example, the soil in your model will not include all of the kinds of particles that can be found in the soil on the real hill.

It is important, however, to design models so they match the real world in the most important ways. Any model you design should match the real-world factors you know are important in managing erosion. For example, you know that the type of material on Earth's surface affects erosion. You also know that the amount of moisture in the material is a factor in erosion. It will be important to decide what real-world conditions should be included in your model. Then design your model to match those real-world conditions as closely as possible.

Design Your Model

Each group will investigate a different erosion-control method by building a model of it on the basketball court site, and then running a simulation. With your group, make decisions together about how to design and build a model of the basketball court site and your erosion-control method. Each group will receive a small container of soil to build a model. The soil in the container is similar to that surrounding the basketball court.

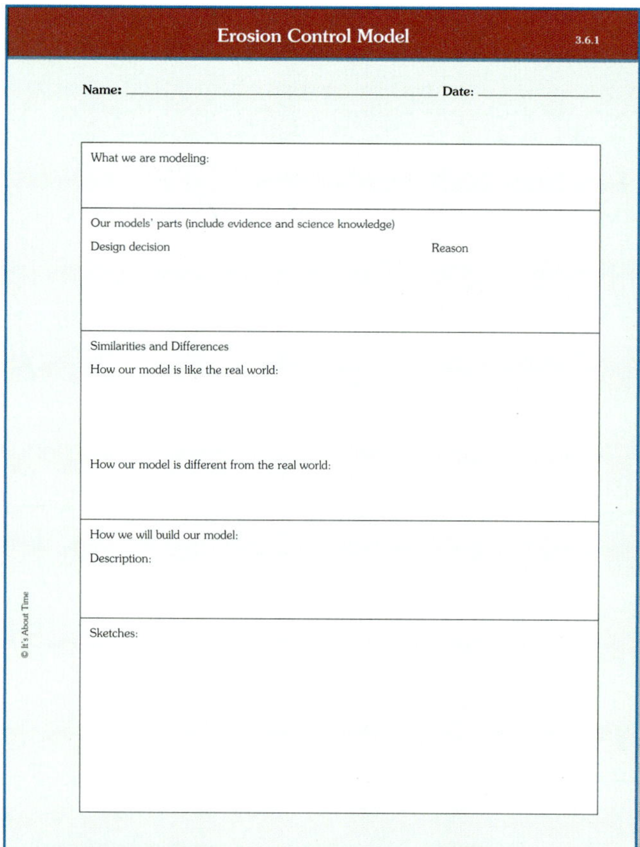
Erosion Control Model 3.6.1

Name: Date:

What we are modeling:

Our models' parts (include evidence and science knowledge)
Design decision Reason

Similarities and Differences
How our model is like the real world:

How our model is different from the real world:

How we will build our model:
Description:

Sketches:

© It's About Time

As you design and build your model, use these questions to guide your decision-making:

- What factors, other than soil type, affect erosion in the real world and should be included in the model?
- How will the scale of the model affect the simulation and your results?
- What constraints, identified earlier in the *Learning Set*, must be addressed in the model?

To model your group's erosion-control method, you will need to make many decisions:

- What materials you will use to model the erosion-control method?
- How will you construct any erosion-control devices you need to create, such as walls?
- Where on the basketball court site will you place those devices?

- What changes, if any, do you need to make to the landscape around the basketball court to use your method?
- How will you attach the devices to the basketball court site?
- What things will you want to find out about your erosion-control method when you run your simulation?

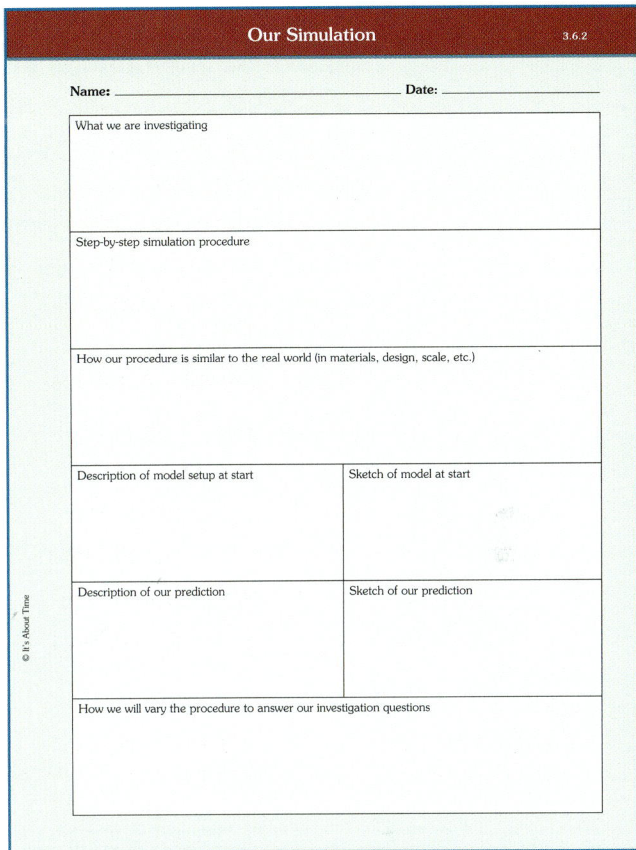
Our Simulation 3.6.2

Name: ______ Date: ______

What we are investigating

Step-by-step simulation procedure

How our procedure is similar to the real world (in materials, design, scale, etc.)

Description of model setup at start | Sketch of model at start

Description of our prediction | Sketch of our prediction

How we will vary the procedure to answer our investigation questions

© It's About Time

Record all of your design decisions, evidence for them, and answers to the questions above in the appropriate places on an *Erosion Control Model* page. You will need to carefully consider everything you have been learning as you make each of these decisions. Record each decision and any evidence you have seen or read about in case studies. Record any science knowledge you know about erosion and its control. You should make informed decisions from the evidence you have available and the science knowledge you know. It is also a good idea to keep track of the things you are not sure about. These are the things you will probably want to use your simulation to investigate. For example, you may not know how deep in the ground a retaining wall has to be positioned to be able to hold back the dirt behind it.

Plan Your Simulation

After designing your model, you will need to plan and design your simulation. You will be running your simulations to learn about how to make your erosion-control method work at the basketball court. You have already identified some questions you want to answer about your erosion-control method. You should design your simulations to help you answer those questions. Use *Our Simulation* pages to record your plan.

Remember that your simulations should re-create real-world conditions as closely as possible. Think about such things as how water should be applied to the model and how much water should be applied to the model. One of your decisions might be whether you will simulate a heavy rainfall or a light rain shower. If you want to find out how heavy a rain your simulation-control method can withstand, you might want to begin with a soft shower and make the rain fall harder and harder over time. You will run several simulations using your model. Think about how you want to start each one. How wet should the soil be, for example? How clean will the bottom of the box need to be for you to make good observations about erosion? Think, too, about how you want to start each simulation. Use the guidelines below to help you as you plan your simulations.

Questions

What questions are you investigating and trying to answer with this investigation?

Prediction

For each, what do you think the answer is, and why do you think that?

Procedure

Write detailed instructions for how to carry out your investigation. Include answers to the following questions:

- How will you set up the erosion-control model?
- How will you simulate erosion?
- How will you measure the model's performance and how will you record the data?
- How many trials will you run, and how will you set up your simulation again after each trial?
- What changes to your model will you have to make to answer your questions, and what different simulations will you need to do?
- What difficulties do you think you might run into when you build your model and run your simulations, and how will you address those?

You might need to set up and run your model several different ways to determine where to put your erosion-control devices and how to attach them to the ground. You will need to collect careful data to answer each of the questions you are investigating. There may be some difficulties collecting that data. It is always good to identify difficulties before you investigate so that you will know what to be especially careful about when you carry out your investigation.

Your goal is to determine the effectiveness of your erosion-control method. You can use your results to identify the strengths and weaknesses of the erosion-control method you are investigating. Remember, that everybody will need your results to be able to address the challenge. When you present your model and plan for investigation to the class, make sure you can tell them why you think they can trust your results.

Use one *Our Simulation* page for each of the simulations you plan. Use the hints on the planning page as a guide. Be sure to write enough in each section so you will be able to present your simulation design to the class. The class will want to know that you have thought through all parts of your plan.

Communicate

To help you as you learn to design models and investigations, you will share your plan with the class. Others in the class have designed models and planned investigations to answer questions similar to the ones you are answering. You will probably see differences and similarities across these plans. In the class discussion, compare the plans. Notice similarities and differences. Identify the strengths of each plan. Think about what could be improved in each. Pay special attention to anything that might help improve your plans.

Revise Your Plan

With your group, revise your plans based on the discussion you just had in class. You might want to change the way you will build your model. You might want to be more specific about the way you run your simulations.

What's the Point?

You have just designed an investigation that will allow you to evaluate your erosion-control method for use at the basketball court site. In the past, you probably followed written steps to run an investigation. Here, you are designing an investigation yourself. Your big challenge is to discover how scientists work together to solve problems. One thing scientists do is collect data and use it as evidence. By designing your own investigation, you will experience how scientists do this.

3.7 Investigate

Simulate Erosion Control

Have your teacher check your plan before you conduct any investigation.

It is time to explore the effectiveness of your erosion-control method. You will do this by constructing your model and then simulating natural forces acting on the hill. As you simulate rain or wind acting on your hill you will be able to see how well your erosion-control method works. Remember, your goal is to investigate the effectiveness of your assigned erosion-control method.

Build Your Model

Your first step is to build your model. Use your revised design plan, and build your model the way you specified in your design.

Before running your investigation, you will need to make sure your model is built the way you intended.

- Does it look the way you want it to look? The size and placement of your hill and the basketball court should match the challenge. Your erosion-control device or method should look like those in the real world.
- Check that everything is in the places where you intend them to be.
- Check that everything is put together well. Anything that needs to be movable should be movable. Anything that needs to be held in place should stay in place well. Remember that moist soil holds things differently than dry soil. Make sure the conditions of the soil in your model are what you intended.

If everything is in place, then you are ready to see if your model will hold together in a simulation.

Test Your Model

Unless your simulation needs to begin with dry soil, you can test your model by gently simulating natural conditions. What happens to it when you apply a gentle rain? What happens with a gentle breeze? Remember that your model is a lot smaller than the real-world basketball court site. Take your model's small size into account when you simulate rain. Think about the size of raindrops, direction of rain, and amount of rain as you decide how to apply a gentle rain to your model.

Try it out. Does everything hold together? If your model doesn't hold together in a gentle rain, it won't hold together in a harder rain. If it begins to fall apart, do what you need to do to make your structures stronger.

Run Your Simulation

Now that you have tested your model, you can use it to investigate the effectiveness of your erosion-control method. You have an investigation plan. It is time now to follow it. Record the results of each of your simulations on *Running our Erosion-Control Model* page. Make sure to record all the different kinds of data you specified in your investigation plan.

Your investigation plan may include changes you will make to the model and simulations using each variation. Make sure each *Running our Erosion-Control Model* page you use states which model you are using for your simulation.

You may also find that some of your simulations happen differently than you were expecting. Sometimes what you observe will give you ideas about other modeling and simulation you should do. For example, a retaining wall might fail in a way that gives you a new idea about where to place it on the hill.

In this kind of investigation, you can decide to change the model a little and run another simulation. Keep three important things in mind as you are doing this.

- Make sure every model and simulation is done for the purpose of answering one of your questions.
- Vary only one thing at a time when you change a model. That way, you can compare two instances of modeling to find out what caused different outcomes.
- Record every change you make to the model as you carry out your investigation. You should also record why you are making every change. To be able to draw conclusions, you will need to know what you did and why.

Running Our Erosion-Control Model 3.7.1

Name: ______________________ Date: __________

Description of setup	Sketch of setup
Explain any changes you made to the procedure	
How our procedure is similar to the real world (in design, scale, materials, amount of moisture, etc.)	
Description of what happened	Sketch of what happened
Summary of measurements and observations	
What did we learn? What else do we need to investigate?	
Conclusions and Recommendations	

Recording Your Observations

As you do the investigation, record the results of each simulation on a *Running our Erosion-Control Model* page. These pages have guidelines on them. They will help you with each task you need to do. Look at the guidelines for hints. Use a separate *Running our Erosion-Control Model* page for each iteration.

Analyze Your Data

Review the data from your investigation. Look for any trends in your results. Use them as evidence to support your answers to the following questions:

1. How did you measure the effectiveness of your method? What happened when it was most effective? What happened when it was least effective?
2. Under what conditions was your erosion-control method effective at controlling erosion? How effective was it? Were there any new problems it caused when it effectively controlled erosion from the hill?
3. What evidence from your past investigations and readings did you use to design your model? Did you find that this evidence applied to this particular situation?
4. What challenge constraints affect the use of this erosion-control method?
5. What problems remain?
6. If you had the chance, what else would you investigate about this erosion-control method before deciding to use it at the basketball court site?

Communicate

Investigation Expo

When you finish your investigation, you will share your results with the class in an *Investigation Expo*. In this *Expo*, each group will take turns standing in front of the class and presenting their erosion-control models and investigation results. So others can learn from what you did and use your results when they design their best erosion-control method, you will make a poster that includes your design, the procedure you used to simulate erosion, and the meaning of your results.

In your presentation, you will explain to the class how your erosion-control model worked (or did not work). Include enough details in your presentation so your class understands how well your model worked to prevent erosion. Since each group tested a different erosion-control method, you need each other's information as you work toward the final model you will present to the school board. Answer the questions on the following page in your presentation.

- What were you trying to find out from your investigation? What was your plan for answering those questions?
- What did you do to simulate erosion? Where did you pour the water onto your model? How much water did you pour? How did you make sure the way you poured the water on your model was similar to real rain? What changes, if any, did you have to make to your earlier procedure to answer your questions?
- How well did your erosion-control method work to manage the erosion? What was the path of the running water? How much soil was carried downhill by the water?
- How, exactly, did your erosion-control method work to manage the erosion? Describe exactly what happened in each trial.
- What changes did you make to your design to improve its performance? What evidence did you use to support those changes?
- How good do you think your model was at modeling your erosion-control method? How much do you trust your results?
- What did you learn about making this erosion-control method effective?

You might find out after running your investigation that your method did not work well. If so, you need to make sure your class knows why this is not a good method to use at the basketball court site.

You might have found that you did not construct your model well enough to learn about your erosion-control method. If that happened, be honest about it. Then explain how you might change the design to make it work better the next time.

As you listen to the investigation presentations of other groups, observe how their different erosion-control methods worked. In what ways is each one good at managing erosion at the basketball court site? What new problems does each one cause?

Reflect

Think about what you learned in the *Investigation Expo* and how well your model worked or didn't work. Answer the following questions and be prepared to discuss the answers with your class:

1. How was your erosion-control model supposed to work?
2. Did it work in the way you thought it would?
3. In what ways were you satisfied with the results?
4. In what ways were you dissatisfied with the results?
5. Given what you read in the case studies and what you saw in your investigations, what new things have you learned? What is your evidence?
6. How are the results of your investigation and the results from other groups going to influence your solution to the challenge? Why?

Update the *Project Board*

You have learned a lot now about how different erosion-control methods work or do not work, and why. It is a good time to go back to the *Project Board* and update it. You can add information in the *What are we learning?* column. Be sure to add the evidence that supports what you have learned in the *What is our evidence?* column.

What's the Point?

You have just run a set of simulations to test the effectiveness of a set of erosion-control methods. You were able to find out many things about how to manage erosion at the basketball court site. That is the point of modeling and simulating real-life situations. It would be far too costly to try out all the different erosion-control methods at the actual site of the proposed basketball court. By using models and running simulations, you can find out what works and what does not work before spending time and money on building an erosion-control system that may not solve your problem.

3.8 Recommend

Which Erosion-Control Methods Might be Appropriate for the Basketball-Court Challenge?

You now have some ideas about how well your erosion-control method works and how to make it work effectively. Perhaps you found that it is not possible to make it work effectively. Whatever you found in your investigation, you should now be ready to make a recommendation to your class about the use of your erosion-control method at the basketball court site.

A recommendation is a special kind of claim where you make a statement about what someone should do. The best recommendations also have evidence and science knowledge associated with them. To make sure your recommendations are good ones, and so you can convince your classmates, you will first state your recommendation as a claim. You will identify the evidence and science knowledge that supports it. You will then create an explanation that supports your recommendation.

Create Your Recommendation

With your group, write a recommendation about the use of your erosion-control method at the basketball court. You might want to write your recommendation in a form that suggests what a person can expect in a situation or what they might do. The *Making Recommendations* box has several examples showing how this can be done well.

Be a Scientist

Making Recommendations

A recommendation is a kind of claim that suggests what to do when certain kinds of situations occur. It can have this form:

When some situations occur, ***do*** or ***try*** or ***expect*** something.

For example, if you want to make a recommendation for crossing the street you might say the following:

When you have the right of way, ***expect*** that some cars will not be able to stop in time.

When you have the right of way, **look** both ways to make sure the traffic has stopped.

Recommendations might also begin with "if." For example,

If you have the right of way, and the traffic has stopped, ***then*** you can cross the street.

Support Your Recommendation

Remember that your recommendation should be written so it will help someone else. The school board members will be especially interested in knowing how you came up with your recommendation. Someone reading your recommendation should be able to apply what you have learned about erosion management and your erosion-control method.

The most trustworthy recommendations (ones people will follow) are those supported by evidence and science knowledge. Use the hints on a *Create Your Explanation* page to make your first attempt at supporting your recommendation. Your evidence will come from the experiments you ran earlier and the modeling and simulation you just finished doing. You can also use science knowledge from the cases you read.

After you have identified the evidence and science knowledge that support your recommendation, write an explanation statement to go with your recommendation that would help someone know why they should trust your recommendation.

Communicate

Share Your Recommendation

When you are finished, you will share your recommendation and explanation with the class. Make sure you trust each recommendation your class members present. If you don't understand the explanation that goes with a recommendation, ask questions. You will want to use your classmates' recommendations, and you can do that only if you believe they are trustworthy. So don't be shy about asking questions, do be respectful.

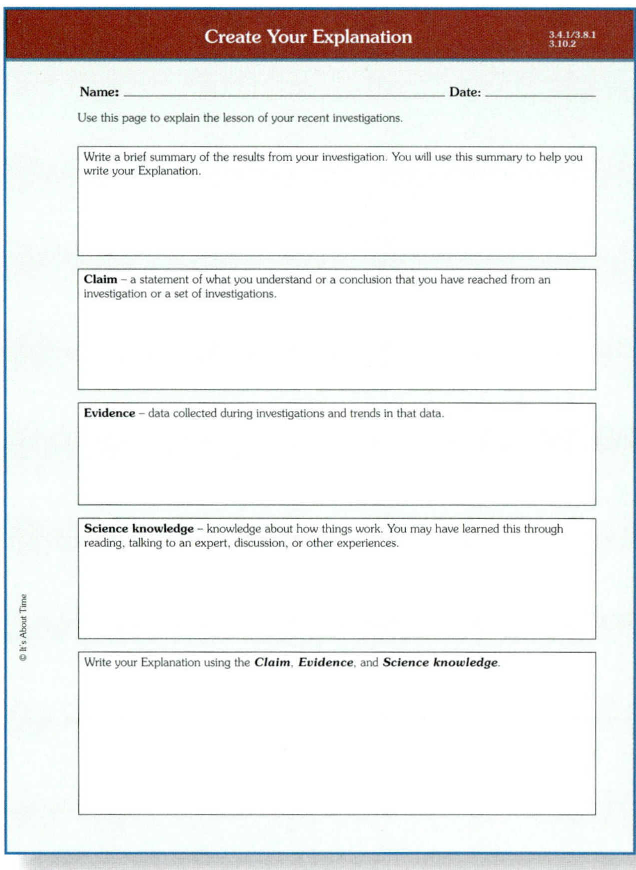

Create Your Explanation 3.4.1/3.8.1 3.10.2

Name: ______ **Date:** ______

Use this page to explain the lesson of your recent investigations.

Write a brief summary of the results from your investigation. You will use this summary to help you write your Explanation.

Claim – a statement of what you understand or a conclusion that you have reached from an investigation or a set of investigations.

Evidence – data collected during investigations and trends in that data.

Science knowledge – knowledge about how things work. You may have learned this through reading, talking to an expert, discussion, or other experiences.

Write your Explanation using the ***Claim***, ***Evidence***, and ***Science knowledge***.

Update the *Project Board*

Your *Big Challenge* is to design an erosion-control method to be used around a proposed basketball court site. The last column on the *Project Board*, *What does it mean for the challenge or question?*, is the place to record recommendations about how to address the challenge. You can now add in this column the recommendations each group wrote and presented for their erosion-control method. In addition, draw lines to the evidence and science knowledge (in the third and fourth columns) that support each of the recommendations. That way, you will be able to keep track of the reasons for each recommendation. When you write up your solution to the challenge, you will want to identify for the school board the evidence you collected that supports your solution.

What's the Point?

Science is about understanding the world around you. Scientists learn about the world by doing investigations. They make claims and provide explanations based on evidence and the science they already know. Then they communicate their results to other scientists. Some claims are in the form of recommendations. A recommendation suggests ways of accomplishing goals. The most trustworthy recommendations are supported by evidence and science knowledge.

Throughout this school year, you will investigate a variety of phenomena. You will apply what you learn to answering *Big Questions* and achieving *Big Challenges*. Many times during each Unit, you will be asked to explain your understandings and knowledge. You will write and share your understanding in explanations. Then you will make recommendations from the explanations you created. To support your recommendations, you will use the same evidence and science knowledge that helps you make claims. As you move through each Unit, you will have many opportunities to edit and improve your explanations and recommendations.

Working as a team, these students have done investigations, made claims, and formulated possible solutions. Now they are collaborating on plans to test their solutions for the Basketball-Court Challenge.

3.9 Plan

Plan Your Basketball-Court Solution

You are about to begin designing and constructing your solution to the *Basketball-Court Challenge*. Your solution can combine as many or as few erosion-control methods as your group feels is necessary. You have learned a lot about how each erosion-control method might work in your situation. You know you have to figure out how to control the water running downhill and the amount of soil and other materials it might carry onto the basketball court.

Using all the knowledge and evidence you gathered throughout this activity, you and your group will create a plan, or blueprint, for your solution. You will present it to your class in a *Plan Briefing*. After that, you will get a chance to build and test a model of your solution and revise your plan.

Design Your Solution

Using everything you have learned, work with your group to design an erosion-control solution for the basketball court. Remember that there are two houses to the sides of the basketball court. The school board says that the two houses and their lots must not be harmed by water or eroding soil.

Some of the erosion-control approaches you have investigated are good at directing water. Others are good at keeping soil in place. Some are not good at all. Your solution should take into account what you have learned about erosion and different erosion-control methods. Remember that the school board is more likely to approve a plan supported by evidence and science knowledge than one that is simply a good idea. As you design your solution, make sure you take into account everything you have been learning.

Using an *Our Design Plan* page, record your design. You will find space for diagrams of your design. You will also find a chart with space for you to list your design decisions and the reasons for each. It has three columns: one for design decisions, one for evidence that led you to make that decision, and one for science knowledge that supports it. You won't have evidence and science for every decision you make, but you need to have one or the other for each.

Most of your design decisions will be based on recommendations made by different groups. Each recommendation has evidence and science knowledge associated with it. If you use the *Project Board* as a resource as you design your solution, it should be easy to fill in the *Evidence* and *Science knowledge* columns of the *Our Design Plan* page.

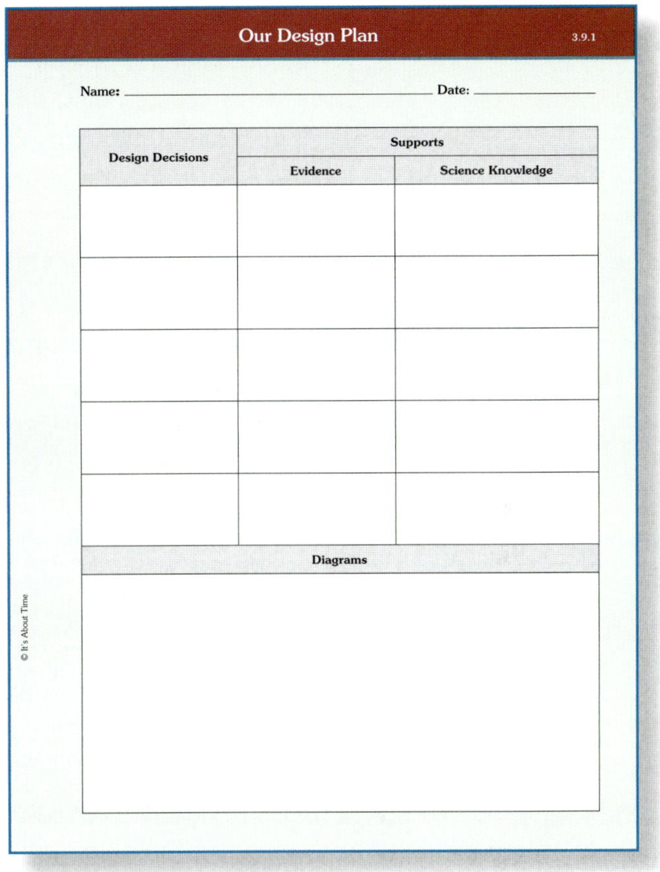
Our Design Plan 3.9.1

Name: Date:

Design Decisions	Supports	
	Evidence	Science Knowledge

Diagrams

© It's About Time

Communicate Your Plan

Plan Briefing

You will present your design plan to your class in a *Plan Briefing*. A *Plan Briefing* is a little like a *Solution Briefing* and a little like an *Investigation Expo*. You will present a plan for a solution. As in an *Investigation Expo*, you will use a poster to organize your presentation.

As in a *Solution Briefing*, a *Plan Briefing* gives you a chance to get advice and suggestions from others. Their advice might help you find a better solution than you could have done by just using the ideas of your group.

You will get good advice from people if they understand why you made each of your decisions. The design plan you recorded on your *Our Design Plan* page should help you prepare your poster for this *Plan Briefing*.

Be a Scientist

Introducing a *Plan Briefing*

Preparing a *Plan Briefing* Poster

A *Plan Briefing* is much like the other presentations you learned to do. In a *Plan Briefing*, you present your design plan. You must present it well enough so your classmates can understand your ideas. They should be able to identify if you have made any mistakes in your reasoning. Then they can provide you with advice before you begin constructing your solution. As a presenter, you will learn the most from a *Plan Briefing* if you can be specific about your design plans and about why you made your design decisions. You will probably want to draw pictures, maybe providing several views. You want everyone to know why you expect your design to achieve the challenge.

The following guidelines will help you as you decide what to present on your poster:

- Your poster should have a detailed drawing with at least one view of your design. You might consider drawing multiple views so the audience can see your design from different angles. It is important that the audience can picture what you are planning to build.
- Parts of the design and any special features should all be labeled. The labels should describe how and why you made each of your design decisions. Show the explanations and recommendations that support your decisions. Convincing others that your design choices are quality ones will convince them that you are making informed decisions backed by scientific evidence.
- Make sure to give credit to groups or students who ran the investigations you used in your design or who gave you ideas that helped your design.
- If another group provided an explanation or evidence that you are using, you should credit them with their assistance in developing your design.

Participating in a *Plan Briefing*

A *Plan Briefing* is similar to an *Investigation Expo* and a *Solution Briefing*. However, this time you will be presenting your design plan. As in other presentation activities, groups will take turns making presentations. After each presentation, the presenting group will take comments and answer questions from the class.

When presenting, be very specific about your design plan and what evidence helped you make your design decisions.

Your presentation should answer the following questions:

- What are the important features of the design?
- What criterion of the challenge will it achieve? What makes your plan the right way to achieve that criterion?

- Are there any problems you foresee with this design?
- What do you predict will happen when you test your design?
- Is there anything you need help with?

As a listener, you will provide the best help if you ask probing questions about the things you do not understand. Be respectful when you point out errors and misconceptions in the reasoning of others. These kinds of conversations will also allow listeners to learn.

For each presentation, if you do not think you understand the answers, make sure to question your classmates. When you ask them to clarify what they are telling you, you can learn more. They can learn, too, by trying to be more precise.

Revise Your Plan

You may have received some good advice from classmates about how to make your design plan better. If so, spend some time with your group doing that. Revise your *Our Design Plan* pages to match your revised plan. Add any evidence and science knowledge that supports your design decisions.

What's the Point?

People have strong opinions about some things. They often assume that their points of view are obvious to others. They think that other people will automatically hold those same views. But often they are surprised. What is obvious to you might look very different to someone else. You probably struggle with this all the time. What seems to work is to describe precisely why you believe what you believe each time you express an opinion.

Evidence that supports your point of view helps other people see your point. When people state their opinions without presenting evidence that justifies them, others are more likely to question their viewpoints. Whenever you need to convince someone of something, or when you are trying to decide between several alternatives, presenting evidence that supports a point of view is critical.

The same is true in convincing yourself that you have made a good decision. You should be able to justify a decision with evidence. Then you will be sure of your decision and more likely to make good decisions.

When you are planning the design of a product or process, it is often useful to hear from others. They can help you see how well your design meets the criteria of the challenge. If you present reasons for the decisions you are making, others can help you identify misconceptions you might have. They can also help you be more confident about your decisions. They can help you judge your decisions based on evidence and knowledge.

An important benefit of a *Plan Briefing* is that teams can learn from each other. A team may have struggled with one aspect of its design. That team may now have good advice for those who have not yet tackled that problem. They, in turn, may benefit from experiences some other team had.

3.10 Build and Test

Build and Test Your Basketball-Court Solution

You planned your best design based on evidence you have available. You presented it to others. You received advice from your classmates. You might have revised your plan based on what your classmates suggested. You are now ready to test your erosion-control solution. You will test it by building a model and then simulating rainfall on the model to see how well it prevents the hill from eroding. You hope to have the erosion-control solution that works the best. If you do, the school board will accept the donated land, the basketball court will be built, and your solution will be implemented!

Build Your Basketball-Court Solution

Work with your group to model and test your solution. You will have the opportunity to revise and test your model two or three times. After you complete your second or third iteration, the class will hold a demonstration and competition. Each group will demonstrate their basketball court solution in front of the class. Each group will also present to the class the changes they made in their design since the *Plan Briefing*. They will explain why they made those changes. When you recommend an erosion-control solution to the school board, you will need to tell them not only how to design the erosion-control solution, but also why you think that design is the best one. On the next pages are some hints for you about how to manage iteration to design your best solution for the basketball court.

Test Your Basketball-Court Solution

Below are some suggestions for testing your designs. Use *Testing My Design* pages to record your work.

Testing Your Designs

In experiments, it is important to run several trials. Then you can be sure your results are consistent. The same is true in testing a design. Each time you test a design, make sure to run enough trials. Choose the number of trials that will allow you to see how it performs. Follow the same procedure each time you test it. Otherwise, you will not know if the design is causing the effects you see or if something you did not control in your procedure is responsible for your results.

Recording Your Work

As you test and revise your design, it will be important to record the results of your tests. You will also need to record the changes you are making. You should record why you are making those changes. This is for several reasons:

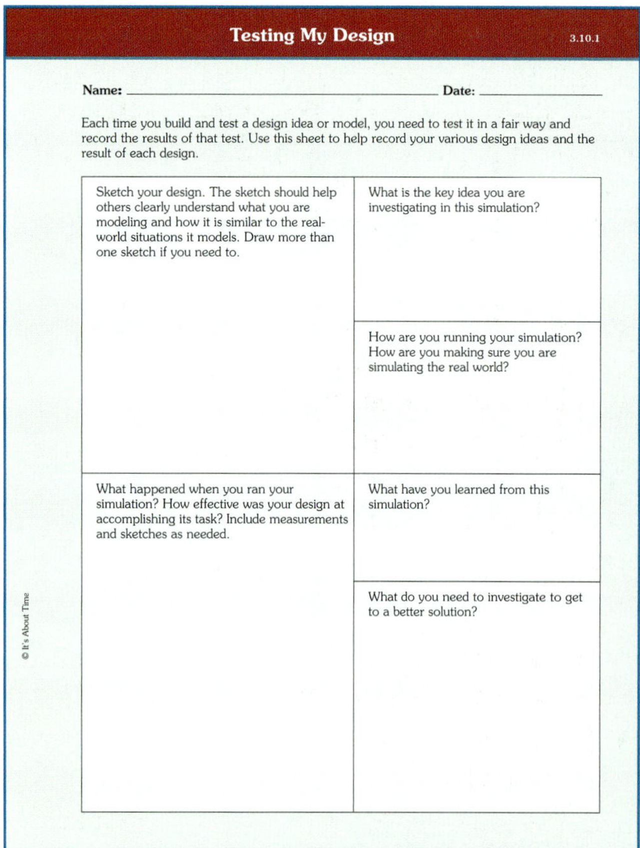

Testing My Design 3.10.1

Name: ______________________ Date: __________

Each time you build and test a design idea or model, you need to test it in a fair way and record the results of that test. Use this sheet to help record your various design ideas and the result of each design.

Sketch your design. The sketch should help others clearly understand what you are modeling and how it is similar to the real-world situations it models. Draw more than one sketch if you need to.	What is the key idea you are investigating in this simulation?
	How are you running your simulation? How are you making sure you are simulating the real world?
What happened when you ran your simulation? How effective was your design at accomplishing its task? Include measurements and sketches as needed.	What have you learned from this simulation?
	What do you need to investigate to get to a better solution?

- Sometimes, what seems like a mistaken approach turns out to work better when some other part of the design is changed.
- You may need to remember what you did and did not test.
- You can use your earlier designs to help teach others.
- By studying your earlier designs, you can learn how your mistakes and successes contributed to your science understanding.

Use *Testing My Design* pages to record your results, changes, and the reasons for changes. Use one page for each iteration.

You can do as many iterations of your erosion-control design as you have time for. However, remember that it is important to run each trial the same way for every design. For example, you must pour the same amount of water onto the model in the same way for every trial. And remember that you need to pour the water to simulate real-world rainfall as well as possible.

Be a Scientist

Iteration

Remember that iteration is a process of making something better over time. That something may be a product, a process, or an understanding. Scientists and engineers iteratively understand new concepts better over time. Scientists iteratively make investigative procedures better over time (as you did in the *Lava Flow* activity).

Engineers iteratively make designs better over time. Each time they test a design, they might find ways to improve it. That is what you are doing now. Sometimes a design does not work as well as the engineer expected. When that happens, the engineer tries to understand why it is not working as well as expected and makes changes based on that analysis. If your design doesn't work as well as you wanted it to, your first feeling may be to throw away those failed plans and begin again. Do not! If you began with a design based on evidence and science, then your solution will probably work well with some changes. Engineers only throw away designs and begin again if the problems are so big that it would be easier or less expensive to begin again.

You saw the power of iteration earlier in this Unit. In the *Build a Boat Challenge*, you improved the design of your boat. You built the boat and tested it. You identified weaknesses and improved your design. Now, with your erosion-control solution, you will again have the opportunity to iteratively enhance your design.

Usually, the best way to iterate is to make one revision at a time. If you make and test one change at a time, then you will know the effect of that change.

Communicate Your Design

Solution Briefing

While you are working on iterating toward a better design, your teacher might have you present your design-in-progress to the class in a short *Solution Briefing*. You will also have a *Solution Briefing* when the time for testing your solutions is over. Recall that a *Solution Briefing* is very much like a *Plan Briefing*. You present your solution for others to comment on.

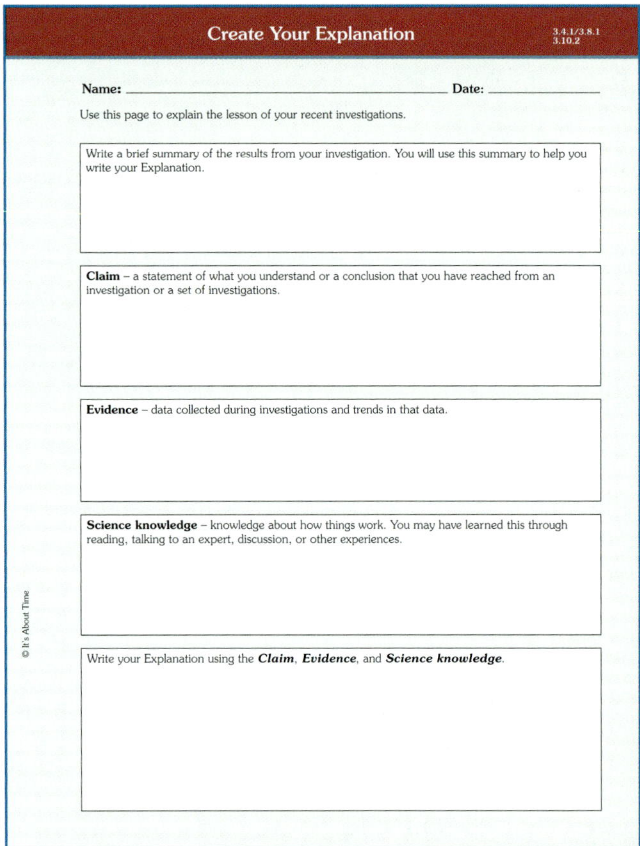

Create Your Explanation 3.4.1/3.8.1 3.10.2

Name: ______ Date: ______

Use this page to explain the lesson of your recent investigations.

Write a brief summary of the results from your investigation. You will use this summary to help you write your Explanation.

Claim – a statement of what you understand or a conclusion that you have reached from an investigation or a set of investigations.

Evidence – data collected during investigations and trends in that data.

Science knowledge – knowledge about how things work. You may have learned this through reading, talking to an expert, discussion, or other experiences.

Write your Explanation using the ***Claim***, ***Evidence***, and ***Science knowledge***.

© It's About Time

An important issue you will have to pay attention to when you engage in these *Solution Briefings* is time. Because you have to break down your designs at the end of each class period, each presentation will have to be short to fit them all in. When your teacher calls a *Solution Briefing,* be prepared to quickly present your progress. Describe the design you are working on. Tell the class how it is different from what you thought you were going to build. Tell them why it is different. Show them what happens when you pour rain on it. Tell them anything you are having trouble with, and ask for advice. Your group's experience may provide valuable lessons for others. If you are having trouble, a *Solution Briefing* is a chance to get help.

Remember, you can learn from attempts that did not work as well as you expected. So do not be shy about presenting what has not worked as well as you expected. You and others can learn from mistakes. Your peers can give you advice about design, construction, and testing.

Reflect

Discuss your erosion-control solution and what you learned from the *Solution Briefing*.

Answer the following questions and be prepared to discuss your answers with the class:

1. How well does your design meet the criteria and constraints of the *Basketball-Court Challenge?*
2. What changes did you make to your design to improve its ability to control erosion?
3. Describe any ideas you got from other groups' designs and presentations or recommendations you used to improve your own design.
4. What changes do you think you still need to make to your design to be more successful?
5. You probably can make some new recommendations about managing erosion, this time about combining erosion-control methods with each other. What new recommendations, if any, should be added to the *Project Board*? Develop your recommendations and their supporting explanations using *Create Your Explanation* pages.

Update the *Project Board*

Based on your experiences combining erosion-control methods with each other, you have derived some new recommendations. Update the *Project Board* with any new recommendations you have and any new evidence you have collected that can help support them.

What's the Point?

You have just built and tested your erosion-control solution. You have run several simulations of your solution and followed the same procedure each time. Each time, you recorded your work and revised your design if it did not work as you expected. You also recorded why you made changes to your design before each new iteration. You presented your design-in-progress to the class in a *Solution Briefing*, asked for advice, and listened to the presentations of others. Based on your experiences combining erosion-control methods, you developed new recommendations and supporting explanations and added them to the *Project Board*. Modeling and simulation is used to test solutions when it would be too dangerous or too expensive to test solutions in the real world. It is always important to be able to predict how a solution will perform before building it. It is also possible to learn how to make a solution better by modeling it.

Address the Big Challenge

Advise the School Board

Your Challenge for this *Learning Set* was to come up with a solution to the erosion problem surrounding a proposed basketball court. You need to make a recommendation to the school board showing how erosion can be controlled without affecting the nearby houses. Your class might want to send one proposed solution to the school board, or you might want to send a set of proposed solutions. You might want to send your set of recommendations. What is important is that you send the school board a package that will convince them to accept the donated land and build the basketball court.

Each group will begin by revising their design one last time based on suggestions made earlier and then presenting their recommendations to the class in a *Plan Showcase*. The class will then decide what solution or set of solutions to send to the school board. The class will work together to produce a good package.

Address the Challenge

You have had experience modeling and simulating your proposed solution to the *Basketball-Court Challenge*. Some of what you proposed earlier worked well, and some of what you proposed did not work as well as you thought it would. It is time now to revise your design one last time and make your recommendation to the school board.

Use an *Our Design Plan* page to record your design and your decisions. Draw as many diagrams of your design as you need. Record your design decisions and the evidence and science knowledge that support them in the appropriate columns. Use the class's *Project Board* to help support your recommendations. The *Project Board* has on it the recommendations your class developed and the evidence and science knowledge used to create those recommendations.

Communicate Your Design

Plan Showcase

Later, you will write a letter to the school board. For now, prepare to present your solution and recommendations to the class in a *Plan Showcase*. A showcase is a presentation that shows off your solution to a challenge or answer to a question. Make a poster that includes the design you recommend and the reasoning behind the decisions you made—the evidence and science knowledge that support them.

You will share your final design in a *Plan Showcase*. Be prepared to share with your class your solution, the reasoning behind it, why you think it is a good solution (even if it does not work 100%), and how you came to your solution. Think of your presentation as practice for presenting to the school board. Remember that your reasoning will be important to the school board as they make their final decision as to whether or not they should accept the donated land and build the basketball court.

Be a Scientist

Introducing a *Showcase*

A *Showcase* is for the purpose of presenting your solution to a challenge or problem or your answer to a question. In a *Plan Showcase*, you showcase your ideas about how to achieve a challenge. You will hold *Solution Showcases* in other Units where you will present actual built solutions. Whichever kind of *Showcase* you participate in, the purpose is to present your solution or answer and help your audience understand what makes it a good solution or answer. In general, you present five things in a *Showcase*:

- the challenge you were addressing or question you were answering, including criteria and constraints;
- your solution;
- the reasoning behind your solution (evidence and science knowledge that support your decisions);
- an analysis of how well your solution addresses the criteria and constraints or answers the question (including what issues you did not address well yet); and

- the history of how you got to your solution, including, for each iteration, the testing you did, your results, your explanations of results, and how that led to your next iteration.

As you listen, it will be important to look at each design carefully. You should ask questions about how the design meets the criteria of the challenge. Be prepared to ask (and answer) questions such as these:

- How well does the design meet the goals of the challenge?
- How did the challenge constraints affect the use or success of this design?
- What problems remain?

Reflect

You will discuss the suggested plans as a class and decide what you should present to the school board. Work in your small group first to identify the strengths and weaknesses of each design. Be prepared to discuss your answers with the class.

- **Which design addresses the challenge the best? Why?**
- **If there is more than one that addresses the challenge well, compare them to each other. What are the strengths and weaknesses of each?**

- Given the strengths and weaknesses of each design, what do you think should be included in the package to the school board?

Use evidence to support your reasoning.

Update the *Project Board*

Now that you have completed the *Basketball-Court Challenge* and decided what to send to the school board, it is time to go back to the *Project Board* for one final update. You will focus mainly on the middle and last columns, filling in what you have learned about the causes of erosion and methods that work well to manage it. Add recommendations to the last column, based on your discussions about solutions and what seemed to work well in controlling erosion at the basketball site.

Advise the School Board

Write a report with your recommendations to the school superintendent, letting her know your best recommendations about the *Basketball-Court Challenge*. Remember that good advice includes recommendations supported with evidence and science knowledge. Your reasoning will be important to the school board as they make their final decision as to whether or not they should accept the donated land and build the basketball court.

Answer the Big Question

Top: *A trio of IDEO designers reviews a proposed concept framework together.* **Middle:** *A project team compares a series of models for a skate park layout.* **Bottom:** *The informal atmosphere of a lounge area acts as a backdrop to a group brainstorm.*

How Do Scientists Work Together to Solve Problems?

You began this Unit with the question: *How do scientists work together to solve problems?* You addressed several small challenges. As you worked on those challenges, you learned about how scientists solve problems. You will now watch a video about real-life designers. You will see the people in the video engaging in activities very much like what you have been doing. You will then think about all the different activities and reasoning you have done during this Unit. Lastly, you will write about what you have learned about doing science and being a scientist.

Watch

IDEO Video

The video you will watch follows a group of designers and engineers at *IDEO*. *IDEO* is an innovation and design company. In the video, *IDEO* designers and engineers face the challenge of designing and building a new kind of shopping cart. These designers and engineers are doing many of the same things that you did. They also use other practices that you did not use. As you watch the video, record the interesting things you see.

After watching the video, answer the questions on the next page. You might want to look at them before you watch the video. Answering these questions should help you answer the *Big Question* of this Unit: *How do scientists work together to solve problems?*

Stop and Think

1. List the criteria and constraints that the design team agreed upon. Which criteria and constraints did the team meet? In your opinion, what other criteria and constraints were not included in the team's discussion?
2. Why did the team split into smaller groups? What did the team hope to accomplish by doing this?
3. What types of investigations did you see the teams doing? What information were the teams trying to collect? Discuss how the information they collected helped the team design a better shopping cart.
4. Why do you think team members' ideas were not being criticized during the initial stages of design?
5. Give at least three examples from the video of how this group of people kept themselves on track to reach their goal on time. (How did they keep the project moving along?)
6. Analyze the team's final product. List three advantages and three disadvantages you see in the new shopping cart.
7. Compare the practices you saw in the video to the practices you used in the classroom. How are they different? How are they the same?
8. Give examples from the video of collaboration and design practices you did not use in the classroom.
9. List two aspects of the *IDEO* work environment you liked. List two aspects you did not like.

10. The IDEO workers have to take on extra responsibilities to maintain their fun, yet productive, work environment. Identify and discuss at least three of these responsibilities.

11. Relate the responsibilities you identified to working with a group in the classroom. Justify your choices using evidence.

Reflect

The following questions review the concepts you learned in this Unit. Your goal was to understand how scientists solve problems. You should start thinking about yourself as a student scientist. The things you are learning about how scientists solve problems will help you solve problems in the classroom and outside of school.

Write a brief answer to each question. Use examples from class to justify your answers. Be prepared to discuss your answers in class.

1. *Teamwork*—Scientists and engineers often work in teams. Think about your teamwork. Record the ways you helped your team during this Unit. What things made working together difficult? What did you learn about working as a team?

2. *Learning from other groups*—What did you learn from other groups? What did you help other groups learn? What does it take to learn from another group or help another group learn? How can you make *Plan* and *Solution Briefings* work better?

3. *Informed decision making*—What is an informed decision? What kinds of informed decisions did you have to make recently? What do you know now about making informed decisions that you did not know before this Unit? What role do results from investigations play in making informed decisions? Provide an example of using results from an investigation to make a decision during this Unit.

4. *Iteration*—Simply trying again is not enough to get to a better solution or understanding. What else do you need to do to be successful? What happens if your design does not work well enough the second time? What if a procedure you are running does not work well enough the second time?

5. *Achieving criteria*—What is a criterion? How do you know which criteria are important? What if you cannot achieve all of them? How did you generate criteria? On which challenges were you able to achieve the whole set of criteria? How did you decide which ones to achieve?

6. *Running experiments and controlling variables*—What does it mean to do a fair test? What is hard about doing a fair test? What happens if you do not control important variables? Some variables are more important to control than others. Why? Use examples from class to illustrate. What did you learn about running experiments successfully that you did not know before? Use examples from class to illustrate your answer.

7. *Modeling and Simulation*—Sometimes a process in the world is too small or too large or too complicated to examine. When scientists need to study the process anyway, they often create models and then run simulations on the models. What modeling did you do in this Unit? You used modeling for three different purposes. What purposes did you use modeling for in this Unit? Simulation means running a model. What kinds of simulation did you do? Modeling and simulation are useful only if the model is similar to the real world in important ways. How did you make sure your models were similar enough to the real world for you to be able to learn from them? How did you make sure you were simulating rainfall in a way that was similar enough to the way it happens in the real world?

8. *Using cases to reason*—Scientists and engineers often have to solve problems that others have confronted before. When scientists and engineers work to solve a problem, they may write up their experience as a case study for others to learn from. They will even write up the case study if their solution did not work. This way, others can learn what not to do to solve the problem. You used case studies several times in this Unit. What are the benefits of using case studies to help you solve a problem?

PBIS

English & Spanish Glossary

A

atom A small particle of matter.

átomo Una pequeña partícula de la materia.

B

bar graph A type of graph that uses either vertical (up and down) bars or horizontal (across) bars to show data. Data can be in words or numbers.

gráfica de barras Un tipo de gráfica que usa ya sea, barras verticales (arriba y abajo) o barras horizontales (a través) para mostrar datos. Los datos pueden ser en la forma de palabras o números.

buoyant force The upward push that keeps an object floating in liquid.

fuerza de flotación El empuje hacia arriba de la fuerza ejercida sobre un objeto que lo mantiene flotando en un líquido.

C

case An example or occurrence of something.

caso Un ejemplo u ocurrencia de algo.

case study An analysis of an example or occurrence of something.

estudio de caso Un análisis de un ejemplo u ocurrencia de algo.

cite To quote something or somebody.

citar Mencionar algo o mencionar lo que dijo alguien.

claim A statement about what a trend means.

reclamo En este caso, una afirmación acerca de lo que significa una tendencia.

collaborate To work together.

colaborar Trabajar juntos.

constraints Factors that limit how you can solve a problem.

limitaciones Factores que limitan tu habilidad para resolver un problema.

control variables In an experiment, the variables kept constant (not changed).

variables de control En un experimento, las variables que se mantienen constantes (que no cambian).

criteria Goals that must be satisfied to successfully achieve a challenge.

criterio Metas que deben ser satisfechas para poder lograr exitosamente un reto.

crystalline The arrangement of atoms in a specific structural pattern.

cristalino La configuración de los átomos en un patrón estructural específico.

D

density The amount of matter in a given amount of space.

densidad En este caso, la cantidad de materia en un espacio determinado.

English & Spanish Glossary

dependent (responding) variables In an experiment, the variables whose values are measured; scientists measure how these variables respond to changes they make in an independent variable.

variable dependiente [de respuesta] Véase variable de respuesta.

deposition In this case, the laying down of soil or other materials in a new location.

deposición El depósito de tierra u otros materiales en una nueva localización.

displace To take the place of.

desplazado: Tomar el lugar de.

drought A long period of dry weather with very little or no rain.

sequía Un período largo de tiempo seco con muy poca o ninguna lluvia.

E

erosion The movement of soil or other materials from one place to another.

erosion El movimiento de tierra u otros materiales de un lugar a otro.

F

fair test Things being compared are tested under the same conditions, and the test matches the question being asked. All variables, aside from the variable you are changing, need to be kept the same.

prueba justa Objetos que están siendo comparados y probados bajo las mismas condiciones, y la prueba corresponde a las preguntas formuladas. Todas las variables, aparte de la variable que estás cambiando, necesitan permanecer iguales.

friction A force that resists motion.
fricción Una fuerza que resiste el movimiento.

G

gravity A pull between two objects. Gravity is the force that holds all objects on Earth.
gravedad Un halón entre dos objetos. La gravedad es la fuerza que sostiene los objetos en la Tierra.

I

igneous One of the three main categories of rock, formed when hot, molten magma or lava cools and becomes solid.
ígneo Una de las tres categorías más importantes de la roca, formada cuando el magma o lava caliente derretida se enfría y se solidifica.

independent (manipulated) variable In an experiment, the variable the scientist intentionally changes.
variable independiente (de manipulación) En un experimento, la variable que el científico cambia intencionalmente.

English & Spanish Glossary

ingenuity Cleverness and originality.
ingenuidad Astucia y originalidad.

inorganic Material that does not come from living things.
inorgánico Material que no proviene de organismos vivientes.

interpret To find the meaning of something.
interpreter Encontrar el significado de algo.

Investigation Expo Presentation of the procedure, results, and interpretations of results of an investigation.
Exposición de Investigación Presentación del procedimiento, resultados, e interpretaciones de los desenlaces de una investigación.

iteration A repetition that attempts to improve on a process or product.

iteración Una repetición que atenta la mejoría de un proceso o producto.

L

lava Melted rock that has reached the surface of Earth.

lava Roca derretida (magma) que ha alcanzado la superficie terrestre.

M

magma Melted rock found beneath Earth's surface.

magma Roca derretida encontrada bajo la superficie terrestre.

matter Anything that has mass and takes up space.

material Cualquier cosa que tiene masa y ocupa espacio.

metamorphic One of the three main categories of rock, formed when already existing rocks are changed by heat, pressure, or chemical action.

metamórfico Una de las tres categorías más importantes de la roca, que se forma cuando rocas ya existentes son cambiadas por medio del calor, la presión o la acción química.

English & Spanish Glossary

mineral A naturally formed, inorganic solid, with a fixed chemical composition and a specific arrangement of atoms in a crystalline pattern.

mineral Un sólido inorgánico de ocurrencia natural con una composición química fija y una configuración específica de átomos en un patrón cristalino.

model A representation of something in the world.

modelo Una representación de algo en el mundo.

molecule The combination of two or more atoms.

molécula La combinación de dos o más átomos.

P

phenomenon An event or detail that can be observed.

fenómeno Un evento o hecho que puede ser observado.

precision How close together measured values are.

precision Cuán cercanos están los valores medidos.

pressure The amount of compression force acting on a substance.
presión La cantidad de fuerza de compresión actuando en una sustancia.

Project Board A space for the class to keep track of progress while working on a project.
Tablón de Proyecto Un espacio para que la clase mantenga un registro del progreso de un proyecto mientras trabajan en él.

R

range The zone between the largest and smallest results.
alcance La zona entre el resultado más grande y el más pequeño.

repeatable When someone follows the reported procedure, they get similar results.
repetible Cuando alguien sigue el procedimiento reportado, obtiene resultados similares.

replicate To run a procedure again and get the same results.
replicar Llevar a cabo un procedimiento nuevamente y obtener los mismos resultados.

English & Spanish Glossary

responding (dependent) variable In an experiment, the variables whose values are measured; scientists measure how these variables respond to changes they make in a manipulated variable.

variable de respuesta En un experimento, las variables cuyos valores son medidos; los científicos miden cómo estas variables responden a cambios que se llevan a cabo en una variable de manipulación.

rocks Naturally formed, non-living solid mass composed of grains of Earth material.

rocas Masa sólida no viviente, de ocurrencia natural, compuesta de granos de materiales de la Tierra.

royalty In this case, a fee paid to the owner of a patent for permission to use it.

regalia En este caso, una tarifa pagada al dueño de una patente para obtener permiso de uso.

S

saturated Unable to hold or contain more; full.

saturado Incapaz de sostener o contener más; lleno.

sediment Solid fragments of inorganic or organic materials that come from rock and are carried and deposited by wind, water, or ice.

sedimento Fragmentos sólidos de materiales inorgánicos u orgánicos que provienen de la roca y son transportados y depositados por el viento, el agua o el hielo.

sedimentary One of the three main categories of rock, formed when eroded and deposited rock particles are compacted or cemented together, or from particles left behind as water.

sedimentaria Una de las tres categorías de rocas, formada cuando las partículas de roca depositada y erosionada son compactadas o pegadas juntas, o por partículas dejadas como agua.

simulate Use a model to imitate or act out real-life situations.

similar Uso de un modelo para imitar o actuar situaciones de la vida real.

soil The loose top layer of Earth's surface, able to support life, consisting of rock and mineral particles mixed with organic matter, air, and soil.

tierra La parte superior suelta de la superficie de la Tierra, capaz de sostener vida, consistente de roca y partículas de minerales mezcladas con materia orgánica, aire y tierra.

standardized The same.

estandarizado Hacer uniforme, regularizar.

T

terracing Cutting a series of raised steps into a slope.

terraza Cortar una serie de escalones elevados en una pendiente.

texture In this case, the size, shape, and arrangement of the grains in rocks.

textura En este caso, el tamaño, la forma y la configuración de los granos en las rocas.

trend A pattern or a tendency.

tendencia Un patrón o un curso.

trial One time through a procedure.

prueba Hacer un procedimiento una vez.

English & Spanish Glossary

V

variable A single factor in an experiment.

variable Un factor único que es probado en un experimento.

variation A spread of data.

variación Una dispersión de datos.

vent An opening on Earth's surface that allows lava, ash, gases, or other volcanic material to escape.

chimenea Una abertura en la superficie de la Tierra que permite el escape de la lava, la ceniza, los gases o cualquier otro material volcánico.

viscous Having a sticky, thick, or gluey consistency; not free flowing.

viscose Tener una consistencia pegajosa, gruesa o engomada; que no fluye libremente.

volcano A place on Earth from which melted rock, ash, gases, and other material can escape from beneath Earth's surface.

volcán Un lugar donde la roca derretida y otros materiales brotan de la superficie terrestre.

volume The amount of space that something takes up.

volumen La cantidad de espacio que ocupa algo.

W

weathering The breaking down of rocks into smaller pieces by natural processes.

erosión (weathering) El rompimiento de una roca en pequeños fragmentos por procesos naturales.

Index

Index

Index

Index

Index

333 North Bedford Road, Mount Kisco, NY 10549
Phone (914) 273-2233 Fax (914) 206-6444
www.IAT.com

Publishing Team

Co-Presidents
Tom Laster and Laurie Kreindler

Director of Product Development
Barbara Zahm, Ph.D.

Managing Editor
Maureen Grassi

Project Development Editor
Ruta Demery

Project Manager
Sarah V. Gruber

Writer
Victoria Willows

Development Editor
Francesca Casella

Assistant Editors, Student Edition
Gail Foreman
Susan Gibian
Nomi Schwartz

Assistant Editors, Teacher's Planning Guide
Danielle Bouchat-Friedman
Kelly Crowley
Edward Denecke
Heide M. Doss
Jake Gillis
Rhonda Gordon

Creative Director
John Nordland

Production/Studio Manager
Robert Schwalb

Production
Sean Campbell
Tommaso Gaetani
Kadi Sarv

Illustrations
Dennis Falcon

Technical Art/ Photo Research
Sean Campbell
Michael Hortens
Marie Killoran

Equipment Kit Developers
Dana Turner
Joseph DeMarco

Safety and Content Reviewers
Edward Robeck
Barbara Speziale

Picture Credits

Page 13	IDEO, Nicholas Zurcher
Page 20	Fotolia/Marzanna Sycerz
Page 21	Jason Harris
Page 40	USGS
Page 43	Rick Busch
Page 44	Rick Busch
Page 55	Pacific Studios, Newport, Oregon; Parke D. Snavely Jr.,USGS
Page 56	Wikipedia; NASA
Page 59	Robert A. Eplett, OES-CA
Page 60	FEMA
Page 62	NOAA
Page 64	USGS
Page 82	FEMA
Page 83—87	City of Austin, Texas
Page 90	Model Builders, Inc.
Page 120	IDEO, Nicholas Zurcher
All other photographs	istockphoto
All Illustrations	Dennis Falcon
Technical Art	Sean Campbell/ Michael Hortens/ Marie Killoran